MOBILITY

ALSO BY JOSPEH M. GIGLIO

Fast Lane to the Future: The Privatization Route
(Hudson Institute, 1996)

MOBILITY

AMERICA'S TRANSPORTATION MESS AND HOW TO FIX IT

Joseph M. Giglio

Washington, D.C.

The views in this book are solely the views of the author. No opinions, statements of fact or conclusions contained in this document can be properly attributed to the Hudson Institute, its staff, its members, its contracted agencies or the other institutions with which the author is affiliated.

Printed in Washington, DC, by Kirby Lithographic Company, Inc.

ISBN: 1-55813-149-3

For information about obtaining additional copies of this or other Hudson Institute publications, contact:

Hudson Institute Fulfillment Center
P.O. Box 1020
Noblesville, IN 46061
Toll free: 888-554-1325
Fax: 317-913-2392

Or, visit Hudson's online bookstore at *www.hudson.org*.

For media and speaking engagement purposes, contact the Hudson Institute at 202-974-2400 or e-mail *info@hudson.org*.

About the Hudson Institute

Hudson Institute is a non-partisan policy research organization dedicated to innovative research and analysis that promotes global security, prosperity, and freedom. We challenge conventional thinking and help manage strategic transitions to the future through interdisciplinary and collaborative studies in defense, international relations, economics, culture, science, technology, and law. Through publications, conferences and policy recommendations, we seek to guide global leaders in government and business.

Since our founding in 1961 by the brilliant futurist Herman Kahn, Hudson's perspective has been uniquely future-oriented and optimistic. Our research has stood the test of time in a world dramatically transformed by the collapse of the Soviet Union, the rise of China, and the advent of radicalism within Islam. Because Hudson sees the complexities within societies, we focus on the often-overlooked interplay among culture, demography, technology, markets, and political leadership.Our broad-based approach has, for decades, allowed us to present well-timed recommendations to leaders in government and business.

For more information, visit *www.hudson.org*

For Anne and Alfred

CONTENTS

CONTINUED ...

The race is not always to the swift
nor the battle to the strong—
but that's the way to bet.

DAMON RUNYON
New York City newspaperman
paraphrasing a verse in Ecclesiastes

AUTHOR'S NOTE

Do we really need another book about transportation problems in the United States?

The answer is "YES, IF ..." with the "IF" involving whether there have been recent changes that make things materially different than they were only a few years ago when other books and articles were written.

One obvious change is the realization that transportation vehicles like vans and commercial airliners can be turned into terrorist weapons. This is a wild card with which we still have to come to terms. But we have witnessed other changes as well, and they may be even more significant in the long run.

This book focuses on these changes, the opportunities they offer, and why they should cause us to think about highways, public transit and goods movement systems as components of an integrated national transportation network rather than as a group of separate travel modes. But first, let us review some basics.

- ***Transportation is closely linked to general economic activity.*** This is because economic activity generates transportation demand. The more economic activity there is, the more transportation capacity we need to support it. Conversely, limitations on transportation capacity can impose artificial constraints on the growth rate of economic activity, just as an automobile factory that is too small can limit the number of cars we can produce to less than the number we could sell. **It is increasingly clear that the United States lacks enough transportation capacity in the right places to meet the demands of today's economy, and this shortfall is growing worse.**
- ***Transportation capacity costs money.*** We must build it, maintain it in decent operating condition so that it performs efficiently, restore it to like-new condition after it wears out through normal use,

and expand it to keep pace with the growth of economic activity. Accomplishing this requires financing mechanisms that can provide enough money in the right places. **But the available mechanisms for financing transportation in the United States can no longer do this.**

- ***Transportation is a national issue.*** That is why Adam Smith devoted so much space in *The Wealth of Nations* to discussing the responsibility of national governments to assure an adequate supply of what he termed "public works." **Unfortunately for the United States, our federal government has not developed any serious national policies for transportation since it created the Interstate Highway system in the 1950s.**

A critical point in this list of basics is that ***economic activity generates travel demand.*** This may seem too obvious to mention. But there is an astonishing lack of understanding about its importance among certain environmentalists and other anti-growth types who have come to strongly influence decisions about expanding roadway capacity in many U.S. metropolitan regions. In fact, leading figures in these circles have become so obsessed by a phenomenon they call INDUCED TRAVEL DEMAND that they have brought new road building to a halt in many regions.

Induced Travel Demand is nothing more than the unsurprising ability of new roadway links to generate trips that were not previously made. What road-building opponents fail to recognize is that very few of these new trips come under the heading of purely wasteful joy riding. Virtually all of them have some economic rationale, so their occurrence means an increase in economic activity.

The ability of new roadway links to generate new trips can legitimately be called INDUCED ECONOMIC ACTIVITY. More economic activity should presumably be welcomed by anyone who believes that rising personal incomes and wealth are positive social outcomes (though this view is obviously not shared by those who have adopted a Marie Antoinette vision of society).

Therefore, let us keep in mind a ruling principle that may seem too obvious to mention but clearly isn't:

A society's transportation systems (including its roadways) exist to support its economy. Not vice versa.

Many books about transportation and other non-fiction subjects are structured in rigid linear fashion like a classic British "whodunit" novel. The reader is expected to peruse each page in the order that it appears and not cheat by skipping over the dull parts and moving ahead to the sections that look more interesting.

But savvy 21st-century readers know that this is not the best way to get the most out of a non-fiction book, especially when time is short and there are lots of other books they have to read. Instead, they tend to proceed more or less as follows:

- They quickly scan the Table of Contents to get an idea of what the book is about and what topics it covers.
- They read the Introduction to determine what the author's shtick is concerning the subject and how he has organized the book.
- Then they jump to the Conclusion section (if there is one) to find out where the author comes out on the subject.
- Finally, they decide whether the book is worth spending any more time with. If "YES," they read some (not necessarily all) of the intervening chapters—usually in the order that interests them rather than in the order that they are printed.

This is how I read most non-fiction books, and it has encouraged me to organize this book in a non-traditional fashion to facilitate this kind of reading. The target audience for this book includes educated laymen, policy analysts in government and universities, as well as dedicated academics and transportation professionals who want to know the HOW and WHY as well as the WHAT. To accommodate the needs of these very different audience segments, I have divided the book into two parts that enable you to read it in whatever manner works best for you.

- PART ONE is a once-over-lightly of the subject written in a user-friendly style to make it easily accessible to the general reader. It covers the topics listed in the Table of Contents for the much-longer Part Two, consists of a number of easily digested short sections, and is mainly policy oriented. Quantitative detail is kept to a minimum and is usually presented in the form of simple graphs that are easy to eyeball. Many of you will find that Part One provides all you need to know about the subject. So don't feel compelled to read beyond this.

- PART TWO covers the same material as Part One. But it covers this material in a series of detailed chapters that are grouped under the key topics. These chapters consist of articles and speeches I've done in the past. (But with some editing by the able and brave Brittany Huckabee to remove stylistic infelicities that should have been red-penciled to begin with. However, off-the-mark predictions, lapses in political judgment and other errors have been left in.) They are organized thematically, and more or less chronologically within the themes, an arrangement which gives them a modicum of coherence. These chapters will be the meat of the book for those of you who are academics or transportation professionals. But some general readers may want to dip into certain chapters to seek more detail about topics that especially interest them. If so, by all means be my guest.

Some notes about my professional career may be helpful because of its influence on the material covered in Part One and the various articles I wrote or co-authored in Part Two.

My career began during the Vietnam War with a stint in the Department of Defense culminating as a systems analyst. Then as now, it was fashionable in many circles to criticize the Defense Department as a gang of bureaucratic bunglers who could never get their socks on straight. This may have been true among top-level political appointees, but it was scarcely the case with most of the military and civilian professionals I had the good fortune to work with and learn from.

Probably the two most important things I learned from these pros related to the complex business of strategy, which I have found to be as important in overhauling the nation's transportation system as it is in fighting wars.

First: Since the future is unknowable, good strategy must hedge against all possibilities and threats. You have to look candidly at all aspects of the external environment and understand their implications before you can formulate meaningful strategy.

Second: While an army may travel on its stomach, it fights wars with the technology it is given. This taught me the importance of understanding the future of technology, its rate of change, and the opportunities it can provide to accomplish new things in new ways.

From the mid-1970s on, I was involved in New York City's recovery from the 1975 fiscal crisis that had posed serious threats to the world's economic stability—first in the city's exceptionally talented government, then in the government of New York State. My particular bailiwick was the tottering healthcare industry, which was critically dependent on the city's cash-starved public hospital system. Our solution had to be two-pronged. We had to improve net cash flows from Medicare, Medicaid and other third party payers, which we did by developing an early version of the Diagnostic Related Grouping Reimbursement system. And we had to reduce costs by negotiating the closure of several hospitals and renegotiating labor contracts that required union workers and healthcare providers to absorb inflation. These experiences gave me an inside view of how public policy is really formulated in the hardball world of clashing political and financial realities. (Bismarck was right about the need for a strong stomach if you want to see how either sausage or public policy is made.)

In the late 1970s I joined a Wall Street firm that had a flourishing securities trading business and wished to expand its investment banking activities. Given the competitive edge enjoyed by established investment banking houses, our best chance for generating a healthy deal flow was to look for new approaches that would address difficult financing problems among potential clients and enable us to make a name for ourselves in this area.

As it turned out, state governments were the most fertile market area for us. For example, one government in a large industrial state was struggling to fund new sewage-treatment plants through a federal grant program that never seemed to provide enough money. Our financing strategy was to use the grants as permanent seed capital for a state revolving loan program instead of the standard approach of using the grants for construction. The state program would make construction loans to sewage-treatment projects, get paid back over time and make new loans with the repayments. We were able to get a federal waiver to make this possible, and the results were so successful that the feds converted the entire grant program into a loan program.

This was the genesis of the State Infrastructure Bank concept, which became an important arrow in the "Innovative Financing" quiver. We were able to generate a nice deal flow for ourselves by applying this concept in a number of states. In effect, we created what amounted to annuities for ourselves by focusing on public policy issues and relationships rather than on traditional one-off financing transactions. It meant we first had to understand the strategic and public policy issues the client was trying to solve, and then design a financing program to address these issues. It also meant we depended on our expanding knowledge of public policy issues (initially drawn from my years in federal, state and local government) to generate opportunities to solicit new clients.

Our success in these activities led to my being named chair of President Reagan's **National Council on Public Works** and the Senate Budget Committee's **Commission on Infrastructure Finance,** which became a crucial element in overhauling transportation financing. I was also a prime accomplice in drafting the liberalized financing provisions in the Intermodal Surface Transportation Efficiency Act (ISTEA) and in the innovative financing concepts pursued by the federal Department of Transportation during the Clinton Administration.

By the mid 1990s, I had retired from investment banking and was spending most of my time on innovative financing as Chairman of Apogee Research, a major consulting firm in the field of transportation and environmental financing. Some of the chapters in Part Two are based on articles that I published in professional journals during these years and reflect my growing understanding of the potential implications of these concepts.

At the same time, I joined the board of the **Intelligent Transportation Society of America** (ITSA) and became its chairman in 2000. This exposed me to the potential of emerging technologies like electronic toll collection to revolutionize transportation. Concurrently, I became a Professor of Corporate Strategy in the College of Business Administration at Northeastern University.

During these years, it became apparent to me that innovative financing suffered from a major flaw. It was failing to provide anything like the NET NEW REVENUES necessary to meet the nation's transportation infrastructure needs. Nor did it address the larger problems of improving operating efficiency, upgrading transportation management and facilitating the use of new technology to emphasize customer service. Instead, innovative financing had become increasingly focused on clever ways to move the same pot of available dollars from one year to another and to design imaginative debt-service schedules that seemed more "painless."

While these innovative financing measures may have been helpful in many ways, they did nothing to increase the basic pot of dollars required to address the four fundamental transportation problems facing the United States:

- Overcoming the pervasive infrastructure deterioration caused by too many years of inadequate funding for maintenance and capital replacement.
- Providing a sensible balance of funds among the various transportation modes so they can work together more effectively.
- Funding the badly needed increases in transportation capacity that the nation's economy demands if it is to achieve its long-term growth potential.
- Enabling the American public and the business community to gain the advantages of a national transportation system for the 21st century that is technologically sophisticated, environmentally beneficial, properly integrated and strongly customer-focused.

The remaining chapters in Part Two are based on more recent articles that reflect my growing awareness of these realities.

Introduction

The most serious problem facing the nation's transportation systems is a choking shortage of money. This is especially critical for the roadway networks that carry the greatest volume of the nation's traffic.

Today's roadway funding depends primarily on motor-vehicle fuel taxes and appropriations from state and local government budgets. But fuel-tax revenues can no longer keep pace with needs because of the self-serving political assumption that it's impossible to "raise taxes" in a society where taxes have become a dirty word. In fact, federal fuel-tax revenue is actually projected to decline in real terms as the nation's motor-vehicle fleet becomes more fuel-efficient. It is safe to say that the federal fuel tax is like a marriage that dies long before divorce papers are filed.

At the same time, state and local government budgets are increasingly burdened with higher funding demands for education ("no child left behind"), fighting street crime ("three strikes and you're out"), beefed-up security against terrorist threats, and a host of other deserving public services. So roadway funding inevitably gets shortchanged. This is easy enough to do since it takes a while for the impact to become apparent.

We were treated to an especially dramatic example of what this impact can be like in late August 2005, when Hurricane Katrina struck New Orleans and caused what is generally regarded as the worst natural disaster in U.S. history. The resulting storm surge overwhelmed the levees protecting this below-sea-level city—leaving nearly 1,000 people dead, 80 percent of the city flooded, whole neighborhoods devastated to the point of being beyond rebuilding, and a repair bill at least a thousand times larger than the original cost of providing New Orleans with the kind of levees that would have prevented all this from happening in the first place.

The U.S. Army Corps of Engineers is responsible for building and maintaining the New Orleans levees. Like most other federal agencies, it is funded through the federal budget—which, like the simple-minded budget of a child's sidewalk lemonade stand, makes no distinction between spending for annual operating expenses and spending for long-term capital improvements.

Each year in its budget requests, the Corps asked for capital funds to improve the levees of New Orleans so they could withstand the storm surges from hurricanes like Katrina. But each year, the White House and Congress cut these requests by up to 50 percent in order to free up funds for farm subsidies, tax cuts for the rich, overseas military adventures and other federal activities they regarded as more important.

To counter any uneasy questions about the danger of a severe hurricane striking New Orleans, agenda-driven federal staffers trained in the nation's elite business schools trotted out artfully worded memos and computer spreadsheets showing that the probability of such an event was too low to worry about. These probability projections relied on something called the **Normal Distribution**, which is the familiar bell-shaped curve found in all standard statistics textbooks. By referring to the Normal Distribution and its accompanying tables, clever analysts can demonstrate that the probability of something bad happening is no greater than an extremely low X percent.

The problem with this kind of analysis is that it implicitly assumes that events in the real world follow the Normal Distribution, or "nearly so." But it turns out that "nearly so" is like being "nearly pregnant." The fact is that many phenomena in the real world don't fit the Normal Distribution. Such phenomena include price changes in securities markets, structural failures in space shuttles and levees, and the incidence of severe hurricanes. The true distribution of these real-world phenomena has higher probabilities of extreme events than the Normal Distribution recognizes. These **Iceberg Risks** (named after the iceberg that sank the *Titanic*—which was also a low probability event according to the Normal Distribution) are what we have to worry about.

But the federal government cared nothing about Iceberg Risks when it came to cutting budget requests for New Orleans' levees. So like a naive gambler in one of Meyer Lansky's original back-alley craps games, it rolled the dice. With disastrous consequences.

The Iceberg Risks associated with the growing backlog of unmet transportation needs may not have the same headline-grabbing consequences as Hurricane Katrina. But their long-term impact could be much worse if they increasingly limit how rapidly the American economy can grow.

For example, roughly $4.4 trillion must be spent on highways and transit systems during the first quarter of the 21st century simply to prevent their already deteriorated condition from getting any worse. Adding in the cost of improving these systems to overcome the effects of past underspending and keep pace with growing demand raises this 25-year cost total to $5.3 trillion. The bar chart below shows the projected cumulative backlog of transportation needs through 2025 and the anticipated gaps based on various estimates of funds expected to be available from existing sources.

CUMULATIVE HIGHWAY AND TRANSIT NEEDS COMPARED TO CUMULATIVE HIGHWAY AND TRANSIT REVENUES, 2005-2025*

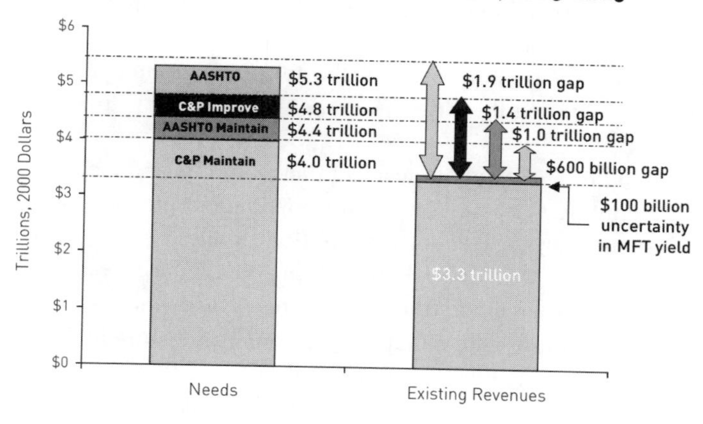

Source: Cambridge Systematics based on FHWA and AASHTO data
*Data is pre-SAFETEA-LU legislation

We can quibble about the actual size of these numbers, just as maritime historians can quibble about the actual size of the iceberg that sank the *Titanic*. But their magnitude is so enormous that it scarcely matters whether the estimates are off by five or ten percent (give or take). What matters are the general proportions of these needs and the Iceberg Risks they imply for the U.S. economy if they are not met. And the longer we wait to address them, the worse they become.

The Last Picture Show

One of the nation's most unusual movie theaters is the Bijou, in an otherwise typical northern California town that we will call Santa Rosita to avoid embarrassing anyone.

Until four years ago, it was no different from any other small-town American movie theater trying to survive on modest ticket sales as the town's last outpost of a vaguely Art Deco Hollywood culture that had largely disappeared elsewhere. But things changed when the elderly owner died of lung cancer and his widow announced that she was going to sell out to a local real-estate developer who planned to convert the Bijou into a combination private gym and sports medicine office building (with each use presumably complementing the other).

For reasons that have never been fully explained but may be obvious, this announcement created a groundswell of dismay throughout the town at the prospect of losing its only traditional movie theater. This dismay reached such proportions that the town's government found itself pressured into buying the Bijou from the owner's widow to keep it open showing movies.

And in a burst of civic enthusiasm that would have done credit to the People's Republic of Santa Monica, the government proceeded to abolish all admissions charges. Henceforth, the Bijou would be open to everyone at no cost "just like a city park or swimming pool," the mayor proclaimed with great pride. Ever since, the Bijou's operating costs have been funded entirely by Santa Rosita's taxpayers through the municipal budget.

Needless to say, this free-movie policy has led to a considerable change in the Bijou's attendance patterns. Virtually no one goes to the movies on weekday afternoons anymore. Even on weekday evenings, the Bijou rarely has more than a handful of moviegoers.

But on weekends when the local schools and most businesses are closed, the picture changes dramatically. The Bijou is full of people eager to enjoy its free movies, with many more waiting patiently in long lines outside for seats to become available. And when the Bijou is playing an especially popular film, these waiting lines begin forming early in the morning well in advance of the noontime opening, reaching such length that Santa Rosita's police department has to assign several of its all-too-few police officers to control the crowds outside the Bijou.

On its face, this seems like a ridiculous way to operate a movie theater. Everywhere else, movie theaters charge admission for access to their seats. They even charge higher ticket prices on weekend evenings when moviegoer demand is at its peak in order to maximize their box-office revenues (which, not so incidentally, tends to spread out demand by encouraging some moviegoers to attend on weekdays when ticket prices are lower).

But the Bijou has no tickets. Access to its seats is free to everyone. That is, free in the sense of not charging any money for seat access. Considerably less than free when you consider the hours moviegoers have to wait in line for seats to become available on high-demand weekends when everyone wants to see free movies.

As ridiculous as this sounds as a system for operating movie theaters, it is exactly the way the United States operates most of its highways. Access to highway lanes is free to all motorists, regardless of the time of day or day of the week and despite the fact that we must pay for access to every other transportation mode.

Free, that is, in the sense of not charging motorists a dollar price for each mile they travel. But scarcely free when we consider the time these motorists have to spend traveling that mile during periods of high demand when bumper-to-bumper traffic reduces average speeds to about ten miles per hour.

Until fairly recently, we could offer the excuse that the logistical problems of directly charging motorists for highway use made the whole idea impractical. Charging for highway use meant toll booths where motorists had to stop and pay out cash from their pockets. And toll booths often meant large toll plazas, which consumed so much land that there could never be enough of them on busy highways to avoid long lines of motorists creeping forward at a snail's pace on the highway. So most highways in the United States had to follow the Bijou Theater's practice of being "free" to all comers.

Technology to the Rescue

Fortunately, new technology is eliminating this excuse. In one form or another, each vehicle can be equipped with a simple electronic gismo that responds to radio signals from roadside transceivers. This enables the roadway's central computer to identify the vehicle, measure the distance it travels along the highway and charge the vehicle owner's computerized account for this distance according to whatever rate per mile is in effect at the time that the trip is made.

This rate can vary depending on the type of vehicle (more for heavy trucks that wear out pavements faster, less for lightweight compact cars), the time of day traveled (more during periods of heavy traffic, less for times when traffic demand is low), the amount of pollution that each vehicle generates per mile of travel, even the actual level of travel demand at any given time. In other words, we can now price access to highways exactly the way we price access to movie seats (except at Santa Rosita's Bijou Theater).

In a world where goods and services aren't available in unlimited quantities, some kind of quantity rationing is inevitable. In the former Leninist nations of Eastern Europe, TIME RATIONING was the standard method. The prices of consumer goods were kept low enough for everyone to afford. But consumers had to spend inordinate amounts of time standing in line to make purchases.

The alternative is PRICE RATIONING. In effect, consumers bid up the price for immediate purchase of a particular good or service until the limited quantity available balances the quantity demanded. This is how the United States rations the supply of most goods and services—with two notable exceptions. One is access to movie seats in Santa Rosita's Bijou Theater. The other is access to virtually all of the nation's roadways. These exceptions use the Leninist concept of time rationing. This favors those who value their time the least and penalizes those who value their time the most (which is not quite the same as saying that the rich and the poor are equally free to sleep under highway overpasses).

There is evidence that motorists respond in positive ways to roadway-lane price rationing in the few locations where it has been tried. They seem willing to make conscious trade-offs between the dollars they spend and the time they save when it comes to making various kinds of trips. The more effective this turns out to be when it is implemented on a sufficiently large scale, the more we can use pricing differences to redistribute travel demand among various roadways and various times of the day. The end result will be to bring the actual daily traffic capacity of existing roadway networks much closer to their theoretical capacity. This means more efficient use of existing lane-miles, thereby reducing the number of additional lane-miles that we need to build.

This follows the principle already used by managers of big-city multiplex movie complexes. They charge higher ticket prices on Saturday night for their theaters playing the most popular movies and lower prices at other times and for their other theaters. The effect is to redistribute seat demand in the direction of maximizing the number of "occupied seat-hours per week," which is a measure of how efficiently a movie theater's manager is using its most important resource (defined as AVAILABLE SEAT-HOURS PER WEEK).

The "pay-as-you-travel" concept for funding highways has a built-in sense of "fairness" that fuel taxes can never enjoy. New technology lets us carry the fairness concept even further by providing discounts to certain population groups such as the elderly, the disabled and the working poor (who are often highly auto-dependent and least able to change their commuting times). By explicitly dedicating the revenue from highway charges to transportation purposes only, we avoid the negative perception dogging all government budgets that "too many of my tax dollars are used to support services that only benefit other people." Pay-as-you-travel means that motorists support the highways they use according to how much they use them.

Finally, we can implement something called PERFORMANCE PRICING on such highways. This means that we guarantee the motorist a certain average speed on the highways that we charge him to use, and we post this speed on variable message signs at each highway entry ramp along with the price per mile. If real-time monitoring shows that the average traffic speed is falling below this level, the rate per mile that the motorist is being charged is automatically reduced (even to zero in the event of accidents or vehicle breakdowns that bring traffic to a virtual halt).

Such a money-back guarantee provides the motorist with confidence that he will enjoy the shorter trip time he is paying for, thereby removing a source of doubt about whether roadway pricing can really deliver what it promises and making the highway that much more attractive. At the same time, the roadway operator has an additional incentive to make sure that the highway is providing truly superior service by maintaining pavements in top condition, rapidly clearing away disabled vehicles and installing the latest technology to speed traffic flow. This is the same principle used by successful private firms that distribute goods and services through the marketplace.

Self-Supporting Roadway Enterprises

The twenty largest U.S. metropolitan regions generate more than half of the nation's gross domestic product even though they contain only about 40 percent of the American population. The density of their urban environments is reflected in the elaborate networks of limited-access highways, secondary arterial roads and local streets on which so much of their economic vitality depends.

But lack of adequate funding for ongoing roadway maintenance, capital replacement of worn-out highway links and overpasses, and capacity expansion to accommodate growing travel demand has left these essential transportation networks increasingly congested and run-down. This poses worrisome Iceberg Risks to future economic growth in these regions and, by implication, in the nation generally.

Under the right circumstances, implementing the "pay-as-you-travel" concept on the limited-access highways in many of these metropolitan regions could generate enough revenue to make their entire roadway networks fully self-supporting in the best free-market sense. This would enable us to end their stepchild-like dependence on the annual government budget-appropriations process for general-fund tax revenues, where they must compete with a host of other taxpayer-supported services that have greater sex appeal. Instead, they would become independent enterprises.

- They would generate enough revenue from user charges to support ongoing operations and maintenance, cover depreciation of their physical assets from normal wear and tear (a cost that American governments traditionally ignore) and fund capital improvements.
- They would be liberated forever from the unrealistic limitations of "penny-per-gallon" excise taxes on fuel, whose revenues fall as automobiles become more fuel-efficient or motorists drive less in response to periodic spikes in gasoline prices.
- They would be subject to the kind of marketplace discipline that encourages them to focus on SERVICE TO CUSTOMERS rather than the traditional engineering-oriented concepts of "operating roadways."
- They would be able to recognize the true significance of life-cycle costs when they make trade-off decisions between capital spending for heavier reconstruction versus annual operating spending for ongoing maintenance.
- Their overall spending levels would be directly linked to the actual travel needs of their customers, as defined by what these customers are willing to pay for different levels of service.

Pairs of metropolitan regions could form cooperative undertakings to have their roadway enterprises fund highways in the low-density rural corridors that connect them. This would assure independence and adequate funding for the external as well as internal roadways on which their economies depend. Equally important, these metropolitan regions could end their present dependence on the federal government for construction grants and other forms of financial aid for roadways. Which means that the federal government (and their state governments as well) would no longer impose fuel taxes in these regions and in the corridors that connect them. Tax revenue would have been replaced by user charges.

The "Common Goods" Issue

Common goods are things like public schools or parks that are collectively owned by society as a whole rather than by any of its individual members. The responsibility for operating and maintaining them is (usually) assigned to society's government, and they are normally supported by tax revenues.

This is the standard pattern for metropolitan roadway systems in the United States. They are built and maintained by a mixture of municipal, county and state governments that fund most of their costs out of general tax revenues. This is often supplemented by "user taxes" levied on the purchase of motor-vehicle fuel—which implies in theory that motorists pay for the roadways they use based on how much use they make of them.

But even in cases where a roadway network is fully supported by fuel taxes, there remains a total disconnect in the minds of motorists between the act of DRIVING on roadways and the act of PAYING for them. This is quite different from the case of commodities that are distributed through the marketplace, where a consumer must buy and pay for some quantity of a commodity before being able to consume it.

The result is an instinctive sense among motorists that "roadways are free." This mistaken view greatly complicates the whole complex of issues relating to how much roadway capacity a society decides that it needs and how that capacity should be managed.

Transportation specialist Joseph F. Coughlin explored this in detail in a paper titled "The Tragedy of the Concrete Commons: Defining Traffic Congestion as a Public Problem." He illustrates the basic problem by using Garrett Hardin's COMMON PASTURE metaphor, in which a society has a publicly owned pasture where local farmers can graze their milk cows without having to pay any user charges. Under these circumstances, each farmer seeks to graze as many cows as he can in the pasture, because each cow he adds to his herd will increase his milk production with no additional feeding cost. So all farmers continue adding cows.

This works well only so long as the total number of cows being grazed remains within the pasture's feeding capacity. But once the farmers exceed this limit, the viability of the pasture begins to break down as the cows consume its grass faster than it can replenish itself with fresh growth. Therefore, the pasture produces less and less nourishment for each cow.

Since each farmer now finds his cows are producing less milk for him to sell, his logical response is to buy still more cows and add them to the overused pasture. When all the farmers do this, the result can only be an increasingly dysfunctional pasture and declining milk production for everyone. In Hardin's words: "Each man is locked into a system that compels him to increase his herd without limit—in a world that is limited. Ruin is the destination towards which all men rush, each pursuing his own best interest in a society that believes in the freedom of the common."

In simple terms, the problem of the overused pasture is rooted in the supply/demand dynamics of classical microeconomics even though the pasture's feeding capacity is not distributed to farmers through a conventional marketplace. But this problem also has sociological overtones because the farmers are acting within the social context of their community. And since the pasture is publicly owned, the problem of overuse raises political issues as well.

The particular "commons" with which Coughlin is concerned in his paper is the nation's roadway system—especially those portions serving metropolitan regions where traffic congestion is severe. Using terms from Hardin's pasture metaphor, he summarizes generally available data that describe the problem and lays out the various solutions being debated to address it.

For example, he notes that some people—such as motorists (or farmers) who believe that their livelihoods depend on free use of the commons (roadway networks or grazing pastures)—propose expanding it to support larger herds (of motor vehicles or cows). Such people believe that the purpose of the commons is to serve the community's economy. Therefore its size should keep pace with economic growth. And since the commons is publicly owned, the cost of expanding it should be paid for out of general tax revenues so that its users can continue to obtain its benefits without having to pay for them directly. Coughlin cites Houston as an example of a metropolitan region that has followed this "expand-the-commons" approach to deal with traffic congestion.

At the other end of the judgment spectrum are those people who insist that the real problem is not too little supply (of roadway lanes or pasture grass) but too much demand (by motorists or farmers). They argue that the time has come to "think green" about the future of public commons in the context of the overall environment. Roadway networks (grazing pastures) are too land-hungry, and environmentally destructive in other ways. Increasing their capacity will only induce further use and therefore add to the environmental damage they cause. Instead, we should begin shifting to more sustainable ways of managing our communities that avoid the need for more roadways (grazing pastures).

Coughlin mentions the San Francisco Bay Area as an example of a metropolitan region that has sought to address traffic congestion by investing in public transportation and implementing various **DEMAND MANAGEMENT** techniques that are designed to reduce auto dependency. (Demand management implies tailoring the size of our economies to transportation capacity rather than vice versa, which is notoriously wrong-headed.)

Then there are those who became enamored of free-market economic theory during their undergraduate years. They suggest that the time has come to begin charging motorists (farmers) user fees—at so much per mile of roadway travel (per hour of grazing time per cow). In this way, each user will pay for the benefits he receives from public commons according to how much use he makes of them. By using a sensible pricing system to ration the use of scarce "commons" resources, each motorist (farmer) will be motivated to make the most efficient use of them.

Also, the income from user fees can cover the cost of expanding such public commons when this becomes necessary rather than taxing everyone in the community to pay for it. This raises the possibility of "privatizing" the commons by selling it (Russian-style?) to the highest or best-connected bidder so that it can be operated in a more business-like fashion.

As the admittedly labored pairings of motorists/farmers and automobiles/cows in the previous paragraphs suggest, Coughlin's grazing pasture metaphor appears to go a long way towards illuminating a broad range of socio-economic questions about why traffic congestion afflicts so many U.S. metropolitan regions.

- It illustrates the inevitable tendency towards overuse of common goods that are perceived to be free.

- It explains why this tendency leads to a condition where supply can never catch up with demand.
- It describes how the widespread availability of free public goods can significantly influence the underlying economics of many private activities that come to depend on them.
- It demonstrates the relative ease with which an entire society can become locked into behavioral patterns that turn out to be counterproductive in the long run.

All things considered, the most sensible solution to the pasture problem may be to charge farmers grazing fees. This immediately confronts farmers with the need to make a series of critical business judgments. Such as how much to spend feeding pasture grass to their cows in order to maximize their revenues from milk production; whether to feed them corn or other grains instead; whether to convert their herds to cows that have higher milk yields but special feeding needs; etc. When all forms of cattle feed are distributed through the marketplace at prices that reflect supply and demand, the business of milk production becomes a more rational undertaking.

The same realities hold true for metropolitan roadway systems. If it becomes possible to directly charge motorists for roadway use (at so much per mile of travel) then the economics of building, managing and using roadways change rapidly—and for the better.

HOUSING-DRIVEN TRAVEL DEMAND

In historical terms, one can argue that many of today's traffic problems can be seen as housing-driven. More specifically, they are driven by the decisions American society made in the late 1940s about how to respond to the critical housing shortages that arose in many metropolitan regions because of abnormally low levels of residential construction during World War II and the years of the Great Depression. Some kind of response was unavoidable due to the waves of returning war veterans who were eager to start families and had access to low-cost home mortgages under the GI Bill.

To meet this need for housing, the nation turned its back on all it had learned about comprehensive national planning during the war. Instead, it chose what seemed like the easier approach of relying almost entirely on "target-of-opportunity" actions by members of the private real-estate industry operating within a largely unregulated free market.

This meant that individual housing developers, acting independently of each other and with little government guidance, optioned inexpensive tracts of vacant land outside existing cities and built clusters of one-family houses on them. By coupling cheap and widely available land with the more efficient construction techniques pioneered by the armed forces during the war, they could build more living space per house at a lower cost than was possible within the city limits.

War veterans, armed with their federally subsidized GI mortgages and with their new wives and children in tow, fell all over themselves in large numbers to buy these brand new "homes of their own" before the paint was even dry—leaving housing developers to pocket their profits, option more tracts of vacant land and build more clusters of houses.

Meanwhile, other real-estate developers saw their targets of opportunity in building retail stores to serve the residents of these new housing tracts. They acquired strips of land along existing local roads when and where they could and erected rows of stores with parking in front, which was necessary because these stores tended not to be within walking distance of anything other than still-undeveloped land.

Local governments responded by laying the miles of new water and sewer mains needed to make the tracts of houses inhabitable. County and state governments followed to build the new roads and highways required to link together these unplanned agglomerations of housing tracts and retail strips and to provide connections to the central cities, where most of the region's jobs were. All of which paved the way for still more target-of-opportunity developments by ambitious real-estate entrepreneurs.

This is how these "organically developed suburbs" began—none of which initially had enough population density for anything other than private automobiles to provide their transportation backbones. And this is how it continued throughout the five remaining decades of the 20th century, with the automobile population inevitably growing too rapidly for new road construction to keep up. It wasn't long before major U.S. metropolitan regions became woven through with the world's most elaborate highway networks—all of which are severely congested during commuting periods, and often at other times as well.

A Different Model

From one perspective, America's postwar approach to dealing with a critical housing shortage can be said to represent a triumph of free-market capitalism in action. It built a great deal of worthwhile housing that has endured far longer and remains in far better condition than most people originally expected. But there are questions about whether the true price of this approach in terms of unanticipated traffic problems may have been too high. Especially given the alternatives.

One of these alternatives can be found in the semi-autonomous Chinese metropolitan region of Hong Kong, which has long been a favorite poster child for academic libertarians because of its reputation for having "the world's freest economy."

But if we zoom back a bit, we find that much of Hong Kong's success as a bastion of free-market capitalism is due to the artful ways in which its highly professional government (run by well-paid career civil servants) has addressed periodic problems that overwhelmed the capabilities of the free market. This enlightened approach to activist government has been a major factor in making Hong Kong's brand of free-market capitalism such a success and giving its nearly seven million residents one of the highest levels of per capita gross domestic product in the developed world.

One of Hong Kong's most serious problems began in the late 1940s when it was inundated by tidal waves of refugees fleeing the Chinese civil war. This caused the region's population to explode and led to a critical housing shortage.

The immediate response of Hong Kong's private real-estate industry to this voracious demand for shelter mainly produced a number of gigantic shantytowns that swarmed like virulent skin rashes over previously vacant hillsides just beyond the region's traditional urban neighborhoods. This was because each individual real-estate developer quickly found that maximum profits came from building crude shacks as quickly as possible on a target-of-opportunity basis and renting them to refugee families (who mostly came from rural backgrounds and had low incomes) for as much as the market would bear.

But after a series of devastating fires swept through several of these shantytowns and left their residents homeless, Hong Kong's metropolitan government found it necessary to step in and become actively involved in housing the region's rapidly growing population. By the early

1970s, the government had effectively taken over full responsibility for meeting the region's NON-DISCRETIONARY need for basic shelter.

In so doing, the government liberated the private real-estate industry from all the profit-constraining social-welfare baggage involved in trying to satisfy the full range of housing demands in an urban society. Instead, private real-estate firms could concentrate exclusively on maximizing their profits by serving niche-market DISCRETIONARY housing demand through a largely unregulated free market—which they have done in very impressive style.

This demonstrates an important economic truth that is too often overlooked. Namely, that ***free-market capitalism works best when it focuses on meeting a society's purely discretionary needs.*** In effect, capitalism requires marketplaces where both buyers and sellers are ***equally free*** to enter into transactions or to turn them down. No market participant should be required to give up this freedom due to circumstances over which he has no control. Otherwise, social support for the role of capitalism quickly erodes.

We saw this happen in the wake of Hurricane Katrina, when temporary reductions in the supply of gasoline caused price spikes at the pump (a normal market-clearing function). This outraged many American motorists and led to dark mutterings about "conspiracies" among the oil companies to manipulate prices, and populist-oriented calls for levying "excess-profits taxes" on these companies. Americans may worship the marketplace as a Wise Goddess when it delivers lower prices for what they regard as non-discretionary commodities like gasoline. But they are quick to condemn it as a Scarlet Woman when it delivers higher prices for these commodities. Hence, the marketplace is generally a "safer" environment for private firms that are in the discretionary commodities business.

But free-market capitalism can only concentrate on meeting discretionary needs if government assumes the responsibility for meeting non-discretionary needs. Even such libertarian economists as Milton Friedman, Friedrich Hayek and Israel Kirzner recognize this (though it may be argued that they tend to take it too much for granted). Hong Kong certainly did when it came to housing its metropolitan population, possibly because of its sophisticated Chinese trading heritage.

The result was that roughly half of Hong Kong's residents now live in some form of government-built housing, where monthly payments by rental tenants are pegged at an extremely modest average of 9 percent of their family incomes (which leaves more income available to support local retail sales, and also eases pressure on employers for wage increases).

Almost instinctively, successful governments tend to approach large problems like housing from a top-down perspective. In Hong Kong's case, this macro bias meant that its government saw regional land-use planning as an important key to addressing the housing shortage (a luxury that individual housing developers don't enjoy). So the increasingly large public-housing complexes that the government built in the decades following the 1950s became characterized by high-rise (therefore land-efficient) apartment towers that also contained retail stores, schools and other neighborhood support facilities. In short, these developments were more like self-contained communities than the isolated "dormitory clusters" typical of suburban housing tracts in U.S. metropolitan regions.

This concept of self-contained communities reached full flower in the nine elaborately planned "New Towns" that now house about 40 percent of Hong Kong's seven million residents and are unlike anything ever seen (or even contemplated) in the United States. They are probably best epitomized by the NEW TOWN OF SHA TIN, which stretches for some 5.5 miles on either side of a park-lined urban waterway in a narrow mountain valley and has more residents than the city of Boston. An extensive local bus network links its high-rise, pedestrian-oriented residential neighborhoods to each other and to Sha Tin's retail centers, commercial offices, educational complexes and other community-support facilities. And two modern electrified commuter rail lines running the length of Sha Tin on either side of its waterway spine provide high-frequency access to the rest of Hong Kong.

All this is obviously quite different from the U.S. approach to dealing with housing demand in metropolitan regions. It reflects a commitment to rational land-use planning on a comprehensive regional basis rather than simply turning loose individual real-estate developers to exploit targets of opportunity on their own initiative.

Among other things, Hong Kong's approach has led to a metropolitan structure that is especially favorable to public transportation as a practical alternative to American-style dependence on the private automobile. While Hong Kong is not without its own traffic problems, more than 85 percent of the trips that its residents make each day are by bus, subway, commuter rail and other forms of public transportation. And only some 16 percent of its families own cars.

This may be especially relevant in view of the growing popularity of the NEW URBANISM movement in a number of U.S. metropolitan regions. This movement seeks to replicate in new suburban developments the land-efficient and pedestrian-friendly character of older (and increasingly gentrified) residential communities in traditional cities like New York, Boston and San Francisco.

Hence the emphasis of New Urbanism on building townhouses rather than tract houses, locating retail stores in small clusters that are within walking distance of townhouse neighborhoods rather than in distant malls that can only be reached by automobile, and generally providing a level of community density that makes walking and public transportation practical undertakings.

At its most ambitious, the New Urbanism movement has the potential for gradually restructuring suburban areas in many metropolitan regions into forms that may be more viable for the 21st century—and in the process, changing their character in ways that address some of the true causes of traffic congestion rather than simply nibbling at the edges.

THE SPECTER OF PRIVATIZATION

The possibility of turning publicly owned metropolitan roadway networks into independent enterprises supported by user charges is an intriguing one. Especially if you can insulate revenue sources and pricing strategies from rapacious elected officials. But it inevitably raises the issue of PRIVATIZATION. This is usually defined (or misdefined) as turning over a traditional government operation to the private sector.

We've all heard the horror stories about the "evils of privatization."

Like the one about the municipal government that was facing a serious budget deficit in an election year when any tax increases would have amounted to political suicide.

So it sold off the town's public library system to a local real-estate developer for enough up-front funds to close the budget gap. With the real-estate developer receiving "service payments" in subsequent years to operate the libraries. Plus the implied promise ("wink, wink") of favorable treatment for his future zoning variance requests.

Or the county highway department that entered into a service contract with an eager private construction firm to take over responsibility for maintaining county roads in exchange for fixed monthly payments.

But the construction firm soon found that the costs of roadway maintenance were higher than it had anticipated, so its expected profits turned out to be losses. Which led to cutting corners in maintenance work so that motorists found themselves driving on roads that were in worse condition than before. Followed by emergency contract renegotiations that raised the monthly payments to levels that at least enabled the construction firm to break even, which wiped out the highway department's anticipated cost savings. Eventually ending up with the maintenance contract being terminated by mutual consent. Which left the highway department back at square one, but saddled with a roadway system in such poor condition due to inadequate maintenance that a great deal of expensive and traffic-disrupting reconstruction was unavoidable.

Or the case of a major American city that shut down its municipally owned factories producing asphalt for roadway repaving on the grounds that it was cheaper to buy asphalt from private firms in the open market.

But it turned out that the local private asphalt industry was an organized-crime operation. So once the city's government had abandoned its asphalt factories, the private firms conspired to raise asphalt prices to levels that were several times more costly than the city had been paying to manufacture asphalt as a municipal enterprise.

(The city in question eventually saw the light and reopened its municipal asphalt factories. They now supply its transportation department with better quality asphalt for roadway paving at much lower prices than it was paying the private firms.)

Possibly the world's greatest privatization disaster story is the case of British Railways, which the Tory government of John Major sold off in the 1990s to a hodgepodge of unrelated private companies in an effort to look good for an upcoming national election (and, not so incidentally, to provide large fees for the London investment bankers who arranged debt and equity offerings for the private companies).

These private companies proceeded to turn a safe, reliable, unimaginative, subsidized government service corporation into a total disaster area that left British socialists howling with glee. This example of privatization-at-its-worst culminated with three serious train wrecks that killed 42 people and injured nearly 600, left the United Kingdom's railway system teetering on the edge of dysfunction, plunged the largest of the private rail corporations into a form of bankruptcy that effectively wiped out all its equity capital, and led directly to the election of a politically independent socialist as the new mayor of London.

Readers who want more such privatization horror stories are directed to Robert Kuttner's 1996 book ***Everything For Sale***, which is full of them.

But the examples described by critics like Kuttner tend to involve cases where privatization has been misused—either because of simple ignorance, the hidden agendas of key players or elaborate schemes to rip off the taxpaying public for personal gain. There is certainly no shortage of ideologues who believe that the greatest virtue in life is to get the government out of any activity that a private firm can run at a profit. Or who have a childlike faith in the tooth-fairy mantra that two plus two can really equal five if a smart businessman is running the adding machine.

What Privatization Is Really All About

In the United States, there are two pragmatic reasons why the private sector is often able to deliver goods and services more effectively than the public sector. They involve irrational anomalies that seem to be deeply embedded in American culture.

The **first reason** relates to ***good management***, which is the key to success in any enterprise. Because private firms aren't hamstrung by American civil-service salary constraints, they can pay the going private-sector rate to attract and keep the best managers. All else being equal, it is scarcely unreasonable to expect the best managers that money can buy to achieve better results than managers of lesser caliber who are willing to work for the much lower salaries typical of most government agencies. It is simply a matter of getting what you pay for. And these days, the marketplace for good managers is too rich an arena for governments to compete.

This anomaly arises from the fact that too many Americans cling to the odd notion that there is a large cadre of trust-fund babies who obtain MBAs from prestigious schools as a more interesting alternative to playing polo, are willing to dedicate their professional lives to "public service" as a quasi-volunteer undertaking, and can afford to do this because their financial futures have long since been secured by the fortune-building talents of their grandfathers. So there is no need to pay public-sector managers as well as their private-sector counterparts.

But apart from a few places like New York City (whose overflowing wealth of talent in all categories has given it the most effective large government in the nation—as its response to 9/11 demonstrated), this notion has no relevance for the overwhelming majority of professional managers. They must build the wealth they need to educate their children and provide for retirement entirely on their own. The only practical way they can do this is by selling their management talents to the highest bidders, nearly all of which are private firms. So even if their personal preferences may incline them towards careers in the public sector, they don't have the luxury of following these preferences.

It seems unlikely that American culture will ever accept the logic of paying public-sector managers salaries that are as high as those in the private sector. Therefore, in cases where good management is the key to success in delivering quality services with maximum efficiency, we have little choice but to rely on the private sector.

The **second reason** builds on the same general principle as the first. Private firms have much more ***flexible procurement regulations.*** This enables them to acquire the newest tools and technology and put them to work producing better services at less cost as soon as they prove themselves.

By contrast, government procurement regulations reflect the American cultural bias that it is better to spend a hundred dollars on triple-chrome-plated oversight procedures than to risk letting a single dollar slip away to a supplier who may not truly deserve it. This commitment to a squeaky-clean procurement environment tends to relegate public agencies to the tail end of the line when it comes to implementing new tools and technology. The inevitable result is that their operating costs are higher and their service quality is poorer than it could be.

Experienced managers know that a fair amount of flexibility is necessary in negotiating and administering contracts with suppliers if the results are to pay meaningful dividends. This can often mean a heavy emphasis on interpersonal relationships and the use of sole-source contracts rather than arms-length open competitive bidding. Traditional critics of the public sector like to call this "honest graft," which has a bad smell to much of the American public regardless of its real-world benefits. They fail to realize that what really counts in the last analysis are results, not civics-classroom myths about what constitutes "good government."

These two pragmatic reasons give private firms a head start in producing better services at lower costs. That is why the public sector has no real choice but to exploit the benefits of privatization whenever and wherever the opportunity presents itself.

This is certainly true if we are to harvest in full the benefits to society of converting metropolitan roadway networks into financially independent enterprises that are self-supporting from highway user charges. The managers of these enterprises must understand in their bones this variation of the classic "make or buy" concept. They must seek every reasonable opportunity to outsource the ongoing maintenance of their roadways, the operation of their toll-collection systems and many of the other services they need by entering into contracts with private firms. Simply following the traditional route of trying to staff up to do everything from A to Z on their own risks turning the concept of self-supporting roadway enterprises into little more than quasi-public jobs programs.

But there is still one fact of life that we must confront squarely. Privatization advocates usually try to hide it under the rug, while opponents like to run it up the flag pole with a loud blare of trumpets as proof that "privatization of public services can never work."

This fact of life is, quite candidly, the inevitability that some private contractors are going to cheat. They will cut corners in the name of "efficiency," arrange with their subcontractors to pad bills for supplies in order to prove that they are spending more than was anticipated to perform the services covered by their contracts, band together in secret cartels to parcel out contracts with public agencies and rig prices, even try to bribe key employees of these agencies in elaborate ways in an effort to gain special consideration.

Such chicanery has a long tradition in the rough-and-tumble world of American capitalism and appears to be unavoidable (after all, you don't survive long in business by leaning over backwards to act like a Sunday school teacher). And the scandals involving public agencies to which it can lead are as appetizing as Peter Luger's steaks to the endlessly famished news media, ranking right up there with spectacular murder trials, celebrity gossip, notable sports events and other widely reported irrelevancies that have little real impact on our lives. But if we're going to start acting like grown-up children when it comes to providing the nation with better transportation and other public services in a more cost-efficient manner, we need to put such chicanery in the right perspective and learn how to make it work for us rather than against us. Doing so requires that we accept some basic truths and learn how to exploit them.

First: In the larger scheme of things, the true costs to the public of chicanery by private contractors may not be all that significant. This is especially true if we try to eliminate it with new and elaborate oversight procedures that end up costing more to implement than the chicanery itself. As noted earlier, the bottom line is what really matters. Is the public receiving better service in a more cost-efficient manner from the use of private contractors, even when this involves a certain amount of behind-the-scenes skullduggery? If the answer is "YES," the wisest policy in serving the public is a judicious amount of "looking the other way" (which is, after all, the American Way unless things get too far out of hand).

Second: When private contractors cheat, they provide public-agency managers with certain opportunities to exploit—such as quietly insisting that the contractors in question provide marginal service enhancements beyond what their contracts require and at no additional cost as the price of not going public with a scandal that could result in some unpleasant criminal indictments for these contractors. Needless to say, this requires agency managers who are reasonably savvy in the ways of the real world and a public that is sophisticated enough to be more concerned with results than procedures.

Third: The potential for chicanery can provide a useful ***creative tension*** across the spectrum of public–private relationships that can have beneficial results. For example, this tension can motivate agency managers and public-employee unions to develop more effective and less costly ways to handle in-house certain work tasks that might otherwise be contracted out. In turn, better and more cost-efficient agency performance can motivate private contractors to devise ways of topping it in order to win contracts. In other words, the potential for chicanery can result in healthy competition that leads to better service to the public.

The real world is no Boy Scout camp. It is a vast, restless marketplace where buyers and sellers must exploit every conceivable angle if they wish to maximize their profits. Ultimately, the winner is the public, which is the inevitable (if invisible) third party in every marketplace transaction. But only if the public belongs to a society that is savvy enough to avoid getting hung up on copybook maxims about so-called morality that are best reserved for Fourth of July speeches by incumbent politicians.

As George Bernard Shaw is reputed to have said: "Capitalism and Religion have a long history of bringing out the worst in people." The solution is not to ban either one, but to find clever ways of tricking both into serving the public interest.

How Privatization Can Be Made to Work Effectively

There is a common myth that privatization must necessarily be an either-or proposition. Either government must produce a public service entirely on its own, with its own employees and its own capital facilities, and with everything under its own control—or it must turn the entire responsibility over to the private sector, giving up all control in the process except for a modest amount of arms-length oversight.

But there turns out to be an alternative. A third way, if you will. Or at least a different way. We can appreciate what this means by looking at our example of a metropolitan roadway network that supports itself by charging variable-rate tolls on its highway links and is therefore converted into an independent transportation enterprise.

For a good model of how this can work, we again turn to Hong Kong. As Adam Smith stressed was necessary in ***The Wealth of Nations***, Hong Kong's government has gone out of its way to provide its region with the top-quality infrastructure needed for the private economy to flourish.

In the field of transportation, for example, Hong Kong's metropolitan government has established and provided equity capital for three large commercial corporations to build and operate its subway system, its commuter rail system and its international airport. Each of these corporations earns attractive profits by providing transportation services that are generally regarded as among the best in the world.

More recently (and in the best venture-capitalist tradition), the government has begun offering dividend-paying minority ownership shares in these corporations to private investors through Hong Kong's stock market in order to recover its original equity investments. But only minority shares. Because in Hong Kong, privatization means that private investors should be allowed to participate on a purely passive basis in the commercial successes the government has created.

This is very different from how the meaning of privatization was corrupted in the United Kingdom, when the government of John Major sold off British Rail to the private sector lock, stock and barrel for political reasons, with little regard for the practical issues that had earlier led Margaret Thatcher (Friedrich Hayek's most famous protégé) to conclude that British Rail was not an appropriate candidate for private ownership and operation.

In the case of independent roadway enterprises in U.S. metropolitan regions, we can effectively "run the Hong Kong model backwards" since their transportation facilities already exist. This could provide the following ownership pattern:

- The state government, which owns the region's limited-access highways, would deed them over to the independent roadway enterprise in exchange for ownership shares. The region's local governments would do the same thing with the arterial roadways and local streets, which they own. Since these governments would have provided the bulk of the enterprise's capital assets, they would collectively hold its majority ownership.
- One class of private investors would purchase equity shares in the enterprise mainly because they have large vested interests in assuring better roadway transportation within the region. These investors would be private utility companies, local banks, large retail chains and other firms based in the region whose future revenue growth depends heavily on rising levels of economic activity in the region.
- A second class of private investors would be firms that purchase equity shares in order to be first in line to win contracts to supply certain goods and services to the enterprise. Such "pay-to-play" arrangements are feasible because the enterprise is an independent commercial corporation rather than a government agency. Therefore, it isn't legally bound to award contracts solely on the basis of arms-length competitive bidding.
- The third class of private investors would be individuals and firms that purchase equity shares simply for their dividend income, just as they purchase shares in private utilities. This income could be enhanced with pass-through tax benefits for the deduction of taxable debt interest and capital depreciation if the enterprise (or the portion of it owned by these investors) is structured as a limited partnership.

At first glance, this may seem like nothing more than an over-elaboration of the traditional public authority model that is common in many parts of the United States. But there are important differences.

A public authority is always wholly owned by government (usually a state government), with no opportunity for private investors to participate. Therefore, a public authority is entirely dependent on issuing debt for the new capital it needs to improve and expand its facilities.

But the roadway enterprise described above enjoys significant amounts of equity capital from private investors and can issue more through public offerings or private placements if necessary. In other words, the roadway enterprise has a much broader spectrum of capital markets it can tap to meet its funding needs. In addition, its ability to obtain "patient capital" in the form of investor equity reduces its need to issue debt.

Another Fine Mess You've Gotten Me Into

So far, we have focused on roadways and explored how they can be converted from "free" public goods like municipal parks and Santa Rosita's Bijou Theater that are supported by taxpayers into independent enterprises that are supported by user charges. We have done this in order to clarify the problems and principles that affect all of the nation's transportation modes.

Roadways may be the nation's number-one transportation mode for moving people and goods when measured by the annual volume of trips they serve. But all trips are not alike in terms of their economic significance. Some are clearly more important than others. That is why we must look more closely at public transit systems, rail lines dedicated to moving goods, canals and other inland waterways, ocean shipping ports, and air travel networks.

All transportation modes have more in common than is generally realized. Our failure to exploit this commonality in a rational manner is one of the reasons why so much of the nation's transportation system is characterized by bottlenecks, costly trip time delays and pervasive inefficiency.

So let us look at some of the nation's other transportation modes.

In March 2004, United Parcel Service had to shift back to trailer trucks its hot-package New York-to-Los Angeles containers from the high-speed daily train service it had worked out with the Union Pacific Railroad.

UPS is known in the trade as a "third-party shipper." This means it provides a service of moving goods that belong to other firms. Therefore its fundamental business principle has to be SERVICE TO CUSTOMERS. In other words, all its management decisions are CUSTOMER-DRIVEN rather than FACILITY-DRIVEN.

This is consistent with the landmark dictum articulated by management guru Peter Drucker a generation ago, to the effect that ***the main goal of every enterprise must be to create customers.*** To avoid any confusion, Drucker further stated that earning adequate profits, paying adequate dividends to shareholders and maintaining the market value of equity shares at adequate levels should be regarded not as business goals but as "necessary costs of doing business"—just like employee salaries and payments to suppliers.

To carry out its principle of service to customers, the choice of what transportation mode UPS uses for any particular goods shipment is determined by which mode can offer the shortest door-to-door trip time, given the customer's trade-off decision between speed versus price. Often, this means using two or more modes for different portions of the shipment's trip.

This embrace of what transportation experts call **INTERMODALISM** is what led UPS to develop its high-speed train service with Union Pacific to move its customers' time-sensitive packages across the thousand miles of empty closet space that separates Los Angeles (the nation's second most important metropolitan region) from the rest of the country. All of which worked very well indeed. Until Union Pacific's growing commodity freight business saturated its available track capacity to a point where it could no longer accommodate the special UPS trains. Hence its "Dear John" letter to UPS discontinuing these trains.

From a narrow cost-accounting perspective, this represents a symbolic triumph for the privately owned U.S. freight railroads in their half-century campaign to boost profits by shrinking the extent of their rail infrastructure. Fewer miles of track means lower operating and maintenance costs to be funded out of revenues from moving goods. These savings go right to the individual railroad's bottom line, though their larger cost implications to the rest of the nation are something else again.

Fifty years ago, American railroad companies had more track capacity than they knew what to do with. Double-track (and even four-track) main lines in many places. Paralleling bypass lines only a few miles apart. Alternate routes galore between U.S. cities. To a point where there was never any problem accommodating more trains.

But all these miles of track were very expensive for the railroad companies to maintain. And too often, these track-miles weren't covering their costs with the revenues they could generate.

In response to this (and accelerated by Congressional passage of the **Staggers Rail Act of 1980**), the railroad companies embarked on a massive track-abandonment campaign to achieve their cost-accounting goal of not-quite-enough track capacity. As a result, total track-miles in the nation's rail network for moving goods declined by 41 percent between 1980 and 2000 even while the nation's volume of goods to be moved continued to increase. From purely a bean counter's point of view, it was considered more profitable to turn away some customers because of insufficient track capacity than to be able to accommodate all comers during periods of peak demand.

But despite this considerable shrinkage in track capacity, the railroad industry's average return on invested assets still remains below its average cost of capital. And because the industry's tracks are privately owned, it is not eligible for any meaningful amount of government aid to enhance and expand this capacity (unlike the publicly owned roadways, airways and inland waterways).

Meanwhile, those UPS containers that once moved by train must now move by truck, which adds to traffic congestion on U.S. highways. The accompanying bar chart shows how domestic freight movements by truck significantly outpaced other modes of freight transportation in 2000. This further inconveniences private automobile drivers.

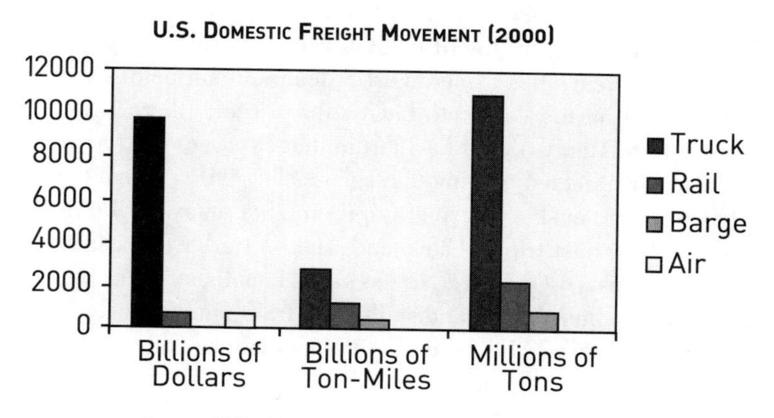

Source: AASHTO, Freight-Rail Bottom Line Report

These motorists were never consulted about this condition. Nor were they compensated in any way for the new inconveniences imposed on them. Not by the Union Pacific, which could claim total innocence because no rational person would insist that it accept more freight business than it could handle. Not by commodity shippers whose growing volume of bulk shipments displaced the UPS trains, since rail is the only affordable way to move these low-value/high-bulk commodities. And not by UPS, which could also claim total innocence since it had been presented with a fait accompli by Union Pacific that left it with no recourse except to shift its time-sensitive package containers to the "free" highways.

But suppose these highways weren't free. Suppose (as we have previously considered) every driver had to pay a fee each time he wished to travel on them. Suppose the size of this fee was determined by how many miles he drives, the time of day he chooses to make his trip, the size and weight of his vehicle, the amount of pollution it generates per mile and the number of other motorists who want to use these highways at the same time.

Now suppose even further that some of the revenue generated by these motorist user fees was used to support more railroad track capacity so that highways could be less crowded with trucks. Simple enough in theory. But in practice, railroads, highways, airlines, ocean shipping ports, inland waterways and transit systems live in totally different worlds as far as transportation funding is concerned.

This disjointed and irrational relationship between various transportation modes similarly affects the travel choices confronting a New York business traveler who must get to a Monday luncheon meeting in downtown Cleveland with an important client.

She can leave her Upper West Side apartment before breakfast and catch a taxi on the corner to LaGuardia Airport (there is still no train service from Manhattan to LaGuardia that bypasses the always-crowded highways in Queens), hoping that she has left early enough to provide an adequate time cushion against unpredictable delays. Like the one that disrupted her last trip to Cleveland when a trailer truck lost its brakes on the Brooklyn-Queens Expressway, demolished four automobiles, killed two of their drivers and stalled the traffic in which she was imprisoned for two hours while she missed her flight to Cleveland and had to postpone her trip to another day.

Or she can fly to Cleveland Sunday evening and check into a downtown hotel. That is, if her company's bean counters will approve the added cost of the hotel plus dinner and breakfast. Not to mention reimbursing her for the extra money she has to pay her day care worker to stay overnight with her daughter.

Is it any wonder she is increasingly tempted by the third alternative? To rent a car Sunday afternoon and drive through the night the five hundred miles to Cleveland, the last portion of which she will have to travel during metro Cleveland's morning rush hour where she contributes to traffic congestion and air pollution. Arriving worn-out and bleary-eyed for her all-important meeting.

Or consider the anxieties dogging the ambitious young real-estate developer who has sunk everything he owns and can borrow into a new residential development around a pristine lake in Rockland County north of New York. He has just spent a Sunday afternoon showing his houses to a young couple from Manhattan expecting their first child. They loved the style of the houses and the prices were well within what they could afford.

But the husband was concerned about the commute to his job in Manhattan. Was there any alternative to a long drive into Midtown? Or a long drive the other way to a commuter rail station where he might (or might not) find parking? Could he seriously think about finding a different job somewhere in northern New Jersey to minimize the time he would have to spend commuting?

The developer's heart sank as he listened to these questions. Because he had heard them all too often from prospective homebuyers, who ended up shaking their heads and walking away. As he stands there alone among his vacant houses, he keeps wondering whatever happened to the long-discussed plan to turn the freight railroad that runs through the nearby town into a commuter rail link to Manhattan. Transit ridership on the nation's bus and rail lines may be up by 24 percent during the last six years. But just as with other transportation modes, the need for new investment to meet this rising transit demand continues to outstrip available funds by significant margins.

Perhaps even more compelling is the plight of the manager of a family-owned apparel chain on Long Island. His main customer base consists of teenage girls, whose mercurial tastes dictate the shortest possible time between his awareness of their lemming-like rush to embrace the latest clothing fad and the arrival in his stores of the new clothes that reflect this fad. For this business manager, the just-in-time delivery concept is more than simply a business-school abstraction. It is a basic reality affecting his bottom line.

In theory, his Hong Kong suppliers can turn out new clothes to his specifications virtually overnight in their Shenzhen factories and pack them into containers that ship out the following evening from the Port of

Hong Kong. In theory, these containers can be transferred from their ships to railroad cars when they arrive at the Port of Los Angeles—then hauled by diesel locomotive up the Alameda Corridor's new rail line to the Union Pacific's vast classification yard in Colton, to be sent across country by train. In theory, these trains can quickly transfer the containers full of clothes to trucks once they reach New York, and the trucks can deliver them to the Long Island apparel stores in time to stock the racks for the coming weekend's shopping blitz.

This theory works well enough in Asia. But it breaks down on this side of the Pacific pond. Like other U.S. ports, the Port of Los Angeles is far behind on the technology curve compared to many foreign ports when it comes to timely unloading of container ships, even though the 14 percent annual increase in freight import volume is outstripping the capacity of their facilities to handle containers efficiently. (The accompanying bar chart illustrates the rise in container traffic nationwide since 1980.)

Trends in U.S. Container Traffic

U.S. Container Traffic in TEUs

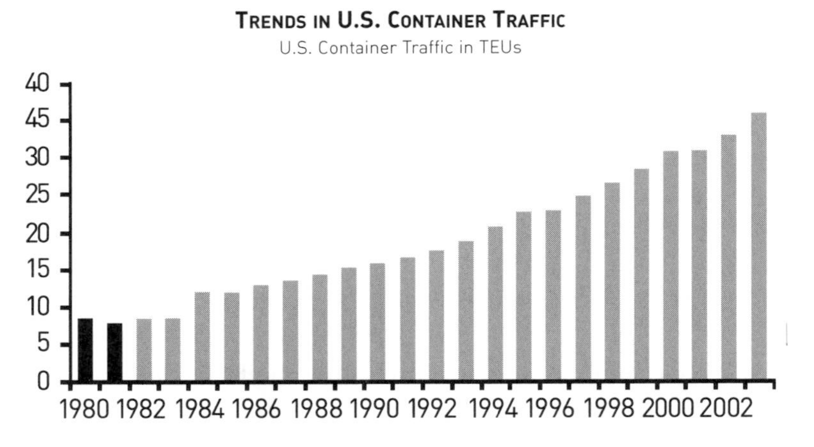

Source: American Association of Port Authorities www.aapa-ports.org

Further, the Alameda Corridor rail line has increasingly become a long storage siding where trains sit parked for too many hours because of congestion in the Colton yard. This is caused by inadequate track capacity on the Union Pacific's lines out of Southern California that prevents the Colton yard from being able to dispatch cross-country trains in a timely fashion.

And when the trains carrying the containers full of clothes finally reach the New York area, they terminate in a railroad yard in northern New Jersey. The trucks onto which the containers are loaded face an hours-long trek up the New Jersey Turnpike, across the George Washington Bridge and the always-congested Cross Bronx Expressway, then down over the Bronx Whitestone Bridge and the traffic-choked Van Wyck Expressway before finally reaching Long Island often too late in the day to make store deliveries.

Is it any wonder that the apparel-chain manager has to build at least an extra week into his shipping schedule from Asia? This often forces him to miss significant changes in the fashion preferences of his teenage customers and end up having to eat obsolete inventory as a result. All because of separate transportation modes that fail to reflect the increasing dependence of today's economy on fast, reliable, fully integrated goods movement from the factory to the customer.

The apparel chain manager may not be a rocket scientist (or even the possessor of an MBA from one of the more fashionable business schools). But he is certain no Neanderthal when it comes to exploiting today's technology to run his business more effectively.

At the end of each weekend, for example, he sifts through the chain's two-day sales records on his laptop computer, whose software quickly shows him with diagrams and tables what customer-buying patterns have been developing for various apparel items. If he sees that there has been a sudden rush to buy jeans with large brass rivets, he can immediately get on the phone with the manager of his main supplier in Hong Kong to talk about this. Since it's already Monday morning in Hong Kong, his supplier can have one of his designers quickly sketch some samples of large-rivet jeans online while they talk so they can all see and agree on the exact appearance of the next jean order to exploit the customer preferences that have surfaced on Long Island that weekend. As soon as the phone call is over, the Hong Kong supplier calls one of his factories in Shenzhen to order production that day of the new jeans for immediate shipment to the United States.

The telephone, computer and software technology that these participants are using is actually quite sophisticated compared to what was available only a few years ago. In fact, it's so sophisticated that the participants aren't even aware of it as they concentrate on doing time-sensitive business together just as if they were meeting face-to-face on the street. This is true even though they are on opposite sides of the world, and only a few hours pass between the time when teenage girls on Long Island indicate their preference for large-rivet jeans and the time when factories in Shenzhen are sewing a fresh order of such jeans. In effect, the participants have learned to *manage* technology in a transparent fashion that eliminates the traditional barriers between different parts of the world, different business firms with different ownerships and managements, and even different positions of the sun in the sky.

Management is really about synthesizing all the functional specialties in an enterprise, its organizational structures, its financing, its technology and ultimately its service to the customer. Management isn't a rigid, by-the-numbers process. It is as messy and complicated as an NBA basketball game and needs to be able to roll with the punches.

But when the containers with those large-rivet jeans reach the Port of Los Angeles, we are back in the slow-moving 1960s where things are still stuck in a Neanderthal time warp. A craft-union crane operator working for a traditional stevedoring company has to move the containers from the ship to railroad flat cars. Then a craft-union locomotive engineer who works for the Union Pacific has to haul the flat cars to Colton (or as near as he can get before his scheduled work hours expire).

When the flat cars finally reach northern New Jersey, another craft-union crane operator working for another company moves the containers onto trucks driven by craft-union drivers who work for a trucking company. They struggle through heavy traffic across two states before the jeans finally reach the Long Island stores whose shopping patterns first gave birth to them. It all unfolds in a very slow and stately fashion, with each company in the transportation chain doing its own thing with only the barest reference to what the others are doing. They are imprisoned by obsolete technology that they still haven't learned to manage with any aplomb.

We can't afford these antiquated minuets any longer. The cost of moving freight the old-fashioned way in terms of higher retail prices, lower business profits, less return on invested assets, all the way back up the line to constrained government tax revenues is becoming a burden that drags down American competitiveness in world markets. This is one of the reasons why so many American firms are moving jobs overseas. And it reflects the all-too-often forgotten truth that the American transportation system exists to support the national economy, not vice versa.

The economy in the United States today is dominated by services and light manufacturing. Not the heavy manufacturing that held sway when the Interstate Highway System was planned and which it was intended to serve. This will be even more so tomorrow, further escalating the already high value of moving people and goods.

Antiquated, commodity-oriented performance measures like ton-miles and vehicle miles matter increasingly less. These are like the "number of units produced" measures that were typically used for state-owned enterprises under the old Leninist system in Eastern Europe and China and that bore no relationship to how well these enterprises were actually serving their economies. Such old-fashioned measures have little relevance for the new economy. They must to be replaced by those involving time and reliability, which come closer to the kind of objective assessments that "after-tax income" provides for business firms.

On July 17, 2003, an 86-year-old man driving alone along a downtown street in the Los Angeles suburb of Santa Monica lost control of his car and plunged into a crowd of pedestrians, killing 10 of them.

This was one of the largest mass killings in Southern California history. If the man had been behind the barrel of a gun, the incident would have been a major news story all over the country for at least several days, just like the Columbine shootings.

But because the man was behind the wheel of an automobile (which routinely kill some 43,000 Americans each year), the whole thing was treated as an "unfortunate accident" and quickly faded away. (The chart below shows the trend in fatalities per 100 million vehicle miles traveled since 1994.) Even so, the incident is significant because of what it tells us about a growing social and economic problem that has major implications for the future of transportation in the United States.

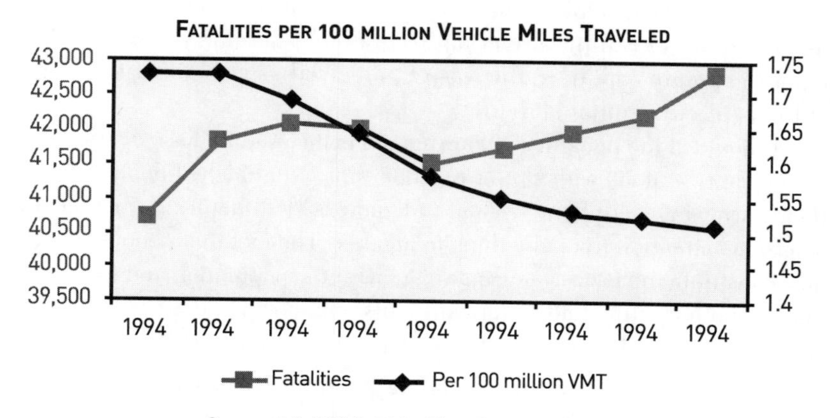

Source: U.S. DOT Fatal Accident Reporting Systems

Throughout most of its century-long infatuation with the automobile, the United States was a young nation with the age distribution of its population heavily skewed towards the 20s and 30s. During this period, the age issue was not related to being too old to drive but of being too young and having to be carted around after school by the equivalent of today's soccer mom. During the decades after World War II when the United States increasingly structured vast areas of its developed landscape around a growing dependence on the automobile, the unspoken assumption was that driving was a natural and desirable means of mobility.

Apart from everything else, there were solid economic reasons for this. For the automobile not only became the most popular passenger transportation mode, it was also the cheapest mode. This was so because it isn't burdened by the high labor costs that periodically help to force other transportation modes like the airlines into Chapter 11.

Instead, ***people drive themselves around.*** They volunteer their transportation-operating labor freely and without question. Asking nothing in return except cheap gasoline and reasonably uncongested roads, which American society was able to provide for a great many years.

In large part, this cost-free provision of transportation operating labor was practical because of the human brain's remarkable capacity for ***multitasking.*** It turned out that a person could drive a car in traffic with surprisingly few accidents while still holding conversations with passengers, thinking about work, scolding his children, eating breakfast or lunch and obsessing about personal problems. Driving a car didn't require one's full-time attention to the exclusion of all else. It could be treated as an incidental activity while the driver concerned himself with other things because of the brain's ability to partition off just enough of its attention to keep the driver out of trouble. Therefore, providing free labor to operate cars in traffic wasn't perceived as something that came at the expense of other activities.

But one of the penalties of growing old is the gradual loss of multitasking capacity—along with slower reaction times, diminished depth perception, narrowing peripheral vision and increased difficulty switching our conscious attention from one thing to another. The fact that people over 65 now constitute the fastest-growing segment of the population and will eventually total one-fifth of all Americans calls into question the basic viability of the auto-dependent society that we have developed since World War II.

- Do we vastly expand the fledgling paratransit systems that provide on-demand chauffeured van service in some communities so that the elderly can be driven where they need to go? If so, who pays for this? The public purse is already struggling to meet the escalating medical costs of the elderly.

- Do we replace the thousands of spread-out, auto-dependent suburbs with more compact townhouse communities that can be served by frequent mini-bus systems and light-rail lines? Organizations embracing the New Urbanism movement appear to have had some success building a few such communities in smaller U.S. metropolitan regions and integrating local stores with residences so that walking-based shopping trips are practical. But this flies in the face of conventional land-use planning and may require some dramatic shifts in public attitudes.
- Do we restructure our whole approach to delivering common consumer goods and services, so that the provider routinely comes to the consumer's home instead of the consumer having to drive to the provider's store?
- Do we move the elderly into assisted-care facilities where they no longer have to worry about driving themselves here and there? They would have to give up their independence in the process so they don't have to be a burden on their children or anyone else but rather to a vague and faceless entity called "society."

No one knows what the practical solutions are when a major portion of an auto-dependent society becomes too old to drive. No one even seems to be thinking about it. We have taken for granted something that has been the norm for at least half a century. Which is that virtually everyone drives.

At what price can Americans remain on the go and romantically obsessed with their cars? That is an issue the nation will have to address.

Consider the choices confronting the elderly couple living in a classic suburban tract house with a large yard. Their vision losses and diminished reaction times make it no longer feasible for them to drive to the local supermarket, to the dentist they've patronized for nearly twenty years, or to the family doctors on whom they depend to treat the growing list of ailments that are inevitable with their advancing age.

Fortunately, they have a married daughter who lives nearby and is able to drive them where they need to go. But the daughter's husband is mulling over an attractive promotion that would require them to relocate to a city across the country.

What is left for the elderly couple in that case? There is no local bus service in their community, where everyone has always had his own car. Taxi services cost more than they can possibly afford. Doctors and dentists have long since stopped making house calls. And supermarkets have always assumed that their customers would come to them.

Must the elderly couple deal with the reality of their loss of personal mobility by moving to an assisted-care facility for people like them if their son-in-law decides to accept the offered promotion? Must they sell their house and give up their long-cherished independence in exchange for life in a combination bedroom/sitting room where everything is presumably done for them?

These examples of the frustrations and economic losses arising from inadequate mobility are increasingly reflected in the broad quantitative measures of national transportation sufficiency. The American Association of State Highway and Transportation Officials (AASHTO) reports that vehicle miles traveled on the nation's roadways grew by some 80 percent during the past twenty years while actual lane capacity increased less than 2 percent (see the chart for the conditions underlying the growth in fuel consumption). According to the Texas Transportation Institute, the resulting increase in roadway congestion translated into 5.7 billion gallons of wasted fuel and the loss of 3.5 billion hours of productivity during 2003 in the nation's 75 most congested areas, for a total congestion price tag of approximately $70 billion.

This squandering of all-too-scarce resources obviously hampers the nation's ability to compete internationally.

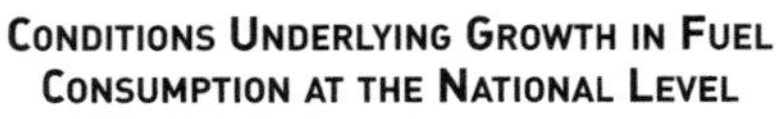
CONDITIONS UNDERLYING GROWTH IN FUEL CONSUMPTION AT THE NATIONAL LEVEL

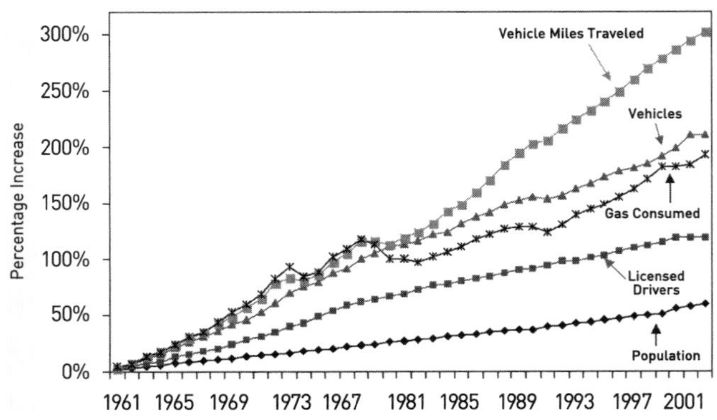

Managing Today's Portfolio of Transportation Modes

By every objective measure, the roadway network is the nation's largest and most important transportation mode. It has the most route miles, moves the most people and carries the most tons of freight. It reigns supreme at the top of the transportation hierarchy because it is the mode that makes the business plans of all other modes economically viable.

Why does it have this capability? First and most obviously, because it connects all the other modes to each other and to the front doors of the people and business firms that generate travel demand.

Second and equally important, because it functions as the "mode of last resort," handling the travel demand that other modes can't or won't accommodate for whatever reasons. That is because access to it is perceived as being "free" (like access to movie seats in Santa Rosita's Bijou Theater). Therefore, it allows the other modes to be distinctly choosy about which travel demand they will serve, when and under what conditions, and in what volume. If the roadway system didn't exist, there is no way that the other modes could function as free enterprises (certainly in the cost accounting sense, and probably in the economic sense as well). They would all have to become like branches of the military, maintained at taxpayer expense to serve "the national purpose" (however defined).

This is because—and it cannot be stressed enough—economic activity inevitably and unavoidably generates travel demand, which must be accommodated at whatever level it requires if economic activity is to flourish. The American tradition has generally been to encourage individual entrepreneurs to stake out niche markets along this demand curve and focus their various transportation service businesses on these markets. Whenever the spectrum of services provided by private entrepreneurs leaves gaps where certain travel demands aren't being accommodated, government steps in to establish public enterprises to fill them. This tradition has worked well enough, but only because it is backstopped by an extensive roadway system that charges nothing to those not served by other travel modes.

If in the name of economic sanity we try to place the roadway system on something like a paying basis by charging motorists for its use and reserving all the revenues generated for roadway-only purposes, we can certainly enhance the system's financial status. But we also risk diminishing its backstopping capability with respect to the other modes. Some travel demand may become too costly to be accommodated as a result. So it will end up not being served and the economic activity it supports can no longer take place.

A more sensible approach (certainly from a broader social and environmental perspective) is to implement roadway pricing in a manner that effectively converts the roadway system into a basic funding source for all transportation modes, while removing the overt price bias that creates the illusion that roadways are cheaper to use than other modes. This would mean that the backstopping function of roadways is increasingly carried out through the financial support they can provide for other modes rather than simply acting as the "price-free" mode of last resort.

By generating revenue that enables the other modes to have increased capacity, the roadway system enables these modes to accommodate some of the travel demand that would otherwise have to move on roadways. This can benefit everyone. It certainly benefits those motorists who enjoy less crowded roads if more freight is moved by rail rather than by truck, more intercity travel takes place by air or train, and more local commuting is done on rail transit lines. It also produces social benefits by accommodating travel demand in a more economically efficient and environmentally friendly manner. Finally, it expands the arena of choices for consumers and producers alike. This enhances the overall productivity of a market-based economy, where more choices are always better than fewer choices.

But some academic economists have a dirty word for this. It's called "cross-subsidization" and is presumed to compromise the inherent vitality of a free-market society, where every enterprise is supposed to "pay its own way." Therefore, the concept of having roadways subsidize other travel modes is a heresy to be avoided.

It is indeed true that there are institutional, administrative and public-acceptance hurdles to be overcome before cross-subsidies can be implemented effectively. These subsidies can take a wide variety of forms. The process of sharing road revenue with transit services in the same corridor where the revenues are collected is quite different from that of using nationwide road revenues to make selected investments in the freight system. There is much work to be done in making these choices work.

But there are two reasons why the academic bias against cross-subsidies is both unrealistic and wrong-headed. The **first reason** concerns the obvious fact that a true marketplace is an open exchange where no enterprise can survive for long without providing benefits to other enterprises in the course of trying to maximize its own profits. This is the whole basis of the barter system, which is an outgrowth of the natural human instinct to trade and is therefore the genetic grandfather of market economics. While it can be argued that today's price-free roadway system is providing benefits to other modes through its backstopping capability, it is doing so in an economically irrational manner that ultimately robs the entire transportation system of the resources it needs to accommodate travel demand in an efficient and socially productive way.

The **second reason** is more down to earth. In turns out that most transactions in the U.S. economy don't take place in what academic theorists would regard as a true marketplace. Rather, they take place within large, vertically integrated corporations where they are subject to top-down central planning in a rigorous command-and-control environment that would presumably do credit to the Leninist ideal. In corporations that are seriously market driven, only the sales function actually generates outside revenues through the marketplace. All other functions (making goods, buying supplies, providing the necessary mix of supporting services, etc.) operate within an integrated framework where cross-subsidies are the norm.

This is what defines a modern market economy. Apart from Mom-and-Pop retail stores that simply buy finished goods for resale, most successful enterprises must create the goods and services they sell on the open market. But the process of creation is so complex as to require numerous specialized internal functions that are effectively cross-subsidized through the external revenues directly generated by the sales function.

The same is true of multi-product corporations like pharmaceutical companies that regard themselves as collections of quasi-independent business portfolios. Each product line functions as a separate business, recording its own revenues and profits and keeping track of what kind of annual return these profits generate on its invested capital.

But not all products are equally profitable in terms of the investment return they generate, and some even run losses in a strict accounting sense. In these cases, does sound portfolio management dictate abandoning these products?

The answer turns out to be "No." This is because some of the company's most profitable new products may depend for their sales volume on customers who regularly purchase older products that are steady sellers but whose profits are modest or even negative. Therefore, these nominally "unsuccessful" products are providing internal subsidies to successful products by delivering customers to them.

Since these cross-subsidies are so important to the company's overall financial health and status as a full-service firm, the more sophisticated forms of portfolio management recognize their value and make appropriate allowances. U.S. auto manufacturers, for example, deliberately produce and sell low-margin vehicles that have high fuel mileage in order to create a window in their federally mandated average-mileage standards for the sale of SUVs and other high-margin vehicles whose fuel mileage is much lower.

Converting the national roadway network into revenue-generating enterprises that can help to fund other travel modes is consistent with this reality. It paves the way for organizing the various travel modes that have too long acted like separate enterprises into a single, integrated transportation entity that can properly position itself to serve the national economy in an increasingly competitive world. Just like a well-managed corporation.

Integrated Surface Transportation

People travel from door to door, and so do goods. In a truly functional sense, they don't travel from passenger station to passenger station, from railroad yard to railroad yard, or even from street corner to street corner. The private automobile is the only travel mode that recognizes this and tries to accommodate it. All other modes hunker down around their separate pieces of the transportation pie, concentrating on what they think they do best to the exclusion of all else, leaving it to the customer to establish his own door-to-door linkages however he wishes.

The great fallacy among transportation providers is that I Run Trains (buses/trucks/ships/highways/etc.). But what they should all be saying is that I Move People and Goods. It is the difference between an inanimate process and a customer. And we must never forget (or perhaps start learning for the first time) that, as Peter Drucker said, creating customers is the most important goal of every enterprise. Without customers, we're just going through the motions. Like one of those bored trust-fund babies who "thinks he'd like to take up sailing."

Hence our first task among transportation providers is to establish an awareness of the customer as the number-one focus of all activities. This requires a new mind-set, a new appreciation of technology and how to manage it, and a new understanding of the many different ways to provide transportation and who fills what niche. (It goes without saying that we must make this awareness financially possible, and we'll get to that in a minute.)

Funding Transportation

Nothing works without financing. Ever since early merchants in medieval Europe persuaded local counting houses to lend them cash so they could buy spices and silks tomorrow from the caravans arriving from China for resale during the next few months to local consumers, money today has been turned into more money tomorrow (certainly enough of the time for the process to have become institutionalized).

But as the economic process becomes more complicated, barriers inevitably develop. Often, these barriers arise as business firms become more complicated, specialized, rigid and distinct. Still, it is possible to speculate (if only in the abstract) about standing in the marketplace, holding up our purse and announcing, "I'm seeking transportation from here to there. Who offers to serve me, how quickly, and for how much?"

In effect, that is the condition we are seeking to restore. But the American surface transportation system has become so ossified that nothing like this can be recreated in simple terms. So we have to improvise.

In the mid-1950s, President Eisenhower appointed his old Army buddy General Lucius Clay to head a commission that would recommend how best to finance construction of the federal Interstate Highway System. General Clay duly reported that federal bonds should be issued to fund these highways, and that the highways should be tolled to generate a revenue stream to retire the bonds and provide for long-term maintenance. This followed the simple and straightforward principle previously adopted by certain states for such highly successful toll roadways as the Pennsylvania Turnpike.

But irrational forces in the Congress twisted this around for political reasons. The end result was:

- No federal bonds or tolls for the Interstate Highway System.
- Pay-as-you-build funding for the highways out of federal appropriations, with modest state participation.
- A federal motor-vehicle fuel tax that was claimed to be adequate for long-term highway maintenance (assuming that inflation remained near zero and the U.S. population didn't really grow).

The bottom line is that the Interstate System got built without tolls and with too little regard for the real impact of life-cycle costs (which is why much of its pavement was under-built to save on construction costs and therefore became a maintenance nightmare later as heavy truck traffic increased). And it has been deteriorating ever since, because fuel taxes don't begin to provide enough revenue to cover the increasingly costly maintenance of its cheaply built pavement in an increasingly inflationary environment where members of Congress seem to believe that any kind of tax increase is a recipe for losing the next election.

So General Clay was right and Congress was wrong. And we are all paying the price in deteriorated and congested highways that are increasingly expensive to maintain while trucks are subsidized, railroad-track mileage shrinks and public transit becomes less competitive.

The best option now is to focus immediate tolling of existing Interstate highways on the most highly congested sections. Together with a reasonable (if temporary) increase in the federal fuel tax to offset past inflation and keep its revenues indexed to future inflation, this can begin to provide the revenue stream needed to prevent highways from deteriorating further, restore some of the capacity we have lost from inadequate maintenance in the past and reconstruct pavements to more rugged standards so they will require less maintenance in the future. Because today's electronic toll-collection technology now makes it possible to retrofit many existing highways so that we can charge for their use on the basis of miles traveled and size of vehicle used, we are in the position of being able to institute fairly quickly a revenue stream that will finally carry us forward rather than backward.

But our ultimate goal must still be the kind of highway pricing system that can provide us with a 21st-century funding source for all forms of surface transportation. No travel volume should be forced onto highways simply because the most logical mode of the transportation system lacks the capacity or the operating dollars needed to accommodate it. In addition, highway users should pay their fair share in a transparent, immediate way that may help to create further opportunities for unsubsidized public transit.

Bringing Technology into the Real World

Critics point out that even if the federal government were to mandate the inclusion of intelligent transportation technologies on all new cars beginning with next year's models (and auto manufacturers were in a position to comply), wouldn't we still have to wait nearly a generation before a large enough proportion of the nation's automobile fleet was equipped with this technology to make a meaningful difference?

Can we wait that long? Or does it mean that relying on such technology isn't practical, and we had better start looking for solutions that aren't burdened by such lengthy lead times?

Fortunately, these problems are already being addressed by the VEHICLE INFRASTRUCTURE INTEGRATION (VII) initiative. This multi-faceted, multi-disciplinary undertaking is managed by a coalition of public and private partners from government (the federal Department of Transportation and ten state DOTs); the automobile industry (BMW, Daimler Chrysler, Ford, General Motors, Honda, Nissan, Toyota and Volkswagen); and key trade associations (AASHTO, the Alliance of Automobile Manufacturers, the Association of International Automobile Manufacturers, the International Bridge, Tunnel and Turnpike Association, the Institute of Transportation Engineers and ITSA). The state DOTS and the auto manufacturers are already involved in the federal DOT's earlier INTELLIGENT VEHICLE INITIATIVE (IVI) program.

The initiative's goal is to enable and deploy a communications infrastructure that supports vehicle-to-infrastructure and vehicle-to-vehicle communications for a variety of safety and mobility applications. In October 1999, the FCC allocated DEDICATED SHORT RANGE COMMUNICATIONS (DSRC) frequencies in the 5.9 GHz band to make such applications possible. Building on this, Mark IV Transportation Technologies Inc. demonstrated an onboard vehicle-to-infrastructure communications system that utilizes the 5.9 GHz band at the November 2005 ITS World Congress in San Francisco.

Based on its schedule for development and exploration, the VII Coalition is expected to announce its final decision for a national rollout of the initiative by 2008. Until then, it will be testing different types of communications systems and their applications, approaches to integration, and competing technologies. The information gained from such R&D will determine what works best and how.

The integration of technology is at least as challenging as the development of the technology itself and could take considerable time. The VII initiative is intent on resolving issues that may seem simple enough in the abstract but turn out to be full of the kind of practical hang-ups that continue to bedevil such federal initiatives as Star Wars.

Keep in mind that integrating multiple large-scale and complex technologies is difficult and expensive. When and if it is completed depends on the "newness" of the technologies and the breadth of the integration. The Space Shuttle, for example, took a very long time and cost billions of dollars. Yet it was successful because the system did not require any breakthrough technologies. The breadth of the integration was also limited—a spacecraft.

But even though people have been working on the Star Wars initiative since the 1970s, the system is not close to being usable. Integrating multiple technologies so that they work is a daunting task. For example, VII may need some technologies, such as computer hardware and software that can perform the real-time, asynchronous analysis of data from thousands of sensors, but that may be years away from working. Star Wars has been limited because no one yet knows how to build computer systems that can track and target thousands of incoming missiles all at once.

Technical feasibility is only one of the areas that the VII Coalition is addressing. Equally important are such areas as institutional relationships, security, privacy, data ownership and product liability. The Coalition will also evaluate economic issues. Who is going to pay for such a communications system? Who will the major users be? What are the capital and operating costs of the roadside infrastructure? Will state DOTs be expected fund these costs out of their existing funds? Given the serious financial problems confronting several of the major auto manufacturers, how much funding can the industry be expected to provide?

An obvious example involves the issue of roadway pricing. Some advocates insist that the technology question is too basic to deserve much attention. The government can simply pass out free signal transponders to all motorists so that the unique ID of each motor vehicle can be read by roadside monitors at each highway on-ramp and off-ramp. That makes it a simple matter for the monitoring system's computers to charge each motorist's account based on the number of miles he actually traveled, the time of day he made his trip and the type of vehicle he used—with the attractive promise of abolishing federal motor-vehicle fuel taxes at the end of the year or so it would presumably take to get such a system up and running. Nice enough in theory. But it ignores some practical realities.

For example, how we deal with vehicles that don't have working transponders? They may belong to visitors from other parts of the country where roadway pricing has not yet been adopted. Or they may be owned by people whose incomes are too low to allow for the credit cards on which most computerized roadway pricing systems are expected to be based. Do we simply prohibit such vehicles from using highways that have pricing systems? Do we provide special cash-payment facilities for these vehicles (which could be logistically complicated and costly)? Or do we build the transponder into each vehicle on the assembly line as part of its original equipment, provide it with a payment account system as part of its registration and somehow work through the complexities of assuring that only vehicles whose payment accounts are current can legally be driven?

The last approach has the advantage of linking onboard pricing technology with other onboard technology that enhances vehicle safety and convenience. But we would probably face a phase-in period of at least several decades before enough suitably equipped vehicles are in operation for roadway pricing to take over the entire burden of funding the maintenance and expansion of the nation's highway system. That's why the most practical approach may be to piggyback on the "smart card" national ID system under discussion at the federal Department of Homeland Security. If a single electronically encoded card can provide secure identification for every adult in the United States, couldn't it also serve as a charge card for paying to use various surface transportation modes—including roadways? But this proposal could further complicate the debate about privacy rights.

These are examples of practical issues that the VII Coalition seeks to resolve in developing a communications infrastructure that could transform the nation's transportation environment by opening up a wide range of safety, mobility and commercial applications.

Near-Term Solutions

Meanwhile, we can no longer put off dealing with the growing backlog of roadway repair and upgrading whose costs have long since outstripped the dollar-generating capacity of the existing roadway funding system. As AASHTO estimates, the existing transportation revenue system provides less than two-thirds of the funds actually needed each year to prevent this backlog from becoming unmanageably worse. So we are clearly on the road to ruin in terms of assuring ourselves of the highway system needed if the nation is to fully realize its economic-growth potential.

But there may be a two-pronged strategy to address this problem. At its heart is a long-term transition to a use-based roadway pricing system that is robust enough to fully meet the nation's roadway funding needs on a consistent basis. The logic of charging drivers for the transportation system they use based on how much they use it and when they use it is beyond dispute. It extends to the roadway system the well-established marketplace principles that have worked so well in delivering other goods and services to the American people. And unlike the free-movies policy that has left the Bijou Theater in Santa Rosita woefully short of the funds it needs to replace its antiquated sound system and overhaul its air conditioning facilities, this approach links how much can be invested in the nation's roadway system with the collective judgment of American motorists about how much transportation they want to consume.

But all this will take time to realize fully. For the near-term interim, the obvious choice is an immediate and temporary increase in the federal fuel tax—but closely coupled with accelerated use of mileage-based roadway fees wherever and whenever available technology makes this feasible. The fuel tax has the virtues of being familiar and accepted, having low transaction costs and being fairly well related to overall road usage. To make it more effective as a revenue tool, the fuel tax should be indexed to keep pace with rising costs driven by growing roadway use and general inflation, and its revenues should be formally lock-boxed for transportation-only purposes to avoid any question of "borrowing" these revenues to fund non-transportation activities.

This toll approach piggybacks on implicit policy changes that are finally surfacing in the White House and Congress. Both branches of government are seeking ways to encourage the growth of roadway pricing in order to prevent the roadway-funding backlog from becoming so unmanageable as to wreck any chances for addressing projected federal budget deficits. But despite the official speeches in favor of roadway pricing, there is still no evidence that the federal government has yet recognized that pricing is simply one element in a larger mix embracing new technology, more savvy management and a sharper focus on motorists as customers rather than taxpayers. Nor is there any awareness that various surface-transportation modes must be seen as components of a single, integrated system instead of continuing to be separate political constituencies.

It is important to note that the proposed fuel-tax increase would truly be temporary. Once enough suitably equipped motor vehicles are operating to place serious roadway pricing on a fast lane to become the main funding source for the nation's roadways, the fuel tax would gradually be phased out. Eventually, only a small residual tax would remain to fund roadways in non-urban states whose low population density may make roadway pricing impractical, and to fund the "special transportation projects" that are so dear to the hearts of many senators and congressmen.

Good-government purists may complain that such special projects are nothing more than pork-barrel giveaways that have become enshrined as part of the vote-trading process (which helps account for the popularity of large transportation bills). While this may be true, it ignores a larger reality. Trading for votes in Congress (not to mention lubricating the process with funding for special projects that the legislators in question consider important) has always been an essential element of American democracy. Stained-glass types may complain that this is little better than bribery. If the use of such a pejorative term makes them feel more noble, so be it. But funding for special projects is a fact of life, and can even be beneficial under the right circumstances. Especially if Congress understands that its control over the use of revenues from the residual fuel tax depends on its cooperation in approving other transportation legislation.

In this context, the avowedly temporary nature of the fuel tax serves as an incentive to impel comprehensive roadway pricing to happen sooner rather than later, with multimodal smart cards helping to accelerate the phase-in by giving drivers a fuel-tax credit when they use toll roads or public transit.

But for roadway-pricing technology to be accepted by American motorists, it must be perceived to deliver superior travel service, with appropriate regard for equity and environmental considerations. Implementation of the money-back guarantee concept should go hand-in-hand with the implementation of roadway pricing from the beginning so that motorists can see for themselves that guaranteed travel-time savings are the flip side of having to pay for access to highways that were once price-free. Making high-polluting vehicles pay more per mile than lower-polluting vehicles is consistent with the principle of using the price system to encourage environmentally positive results. Both concepts come under the heading of managing technology properly, and the management of technology (including expanding the use of smart ID cards to pay for travel on a variety of surface-transportation modes) is at least as important as its development.

Safety Issues

Automobiles are currently killing more than 40,000 people a year in the United States, and this number is rising. This clearly qualifies as a major health epidemic, whose resulting medical care and property damage impose increasing cost burdens on the American public. If such a death toll were attached to an obvious viral disease, the public outcry would be sufficient to force the federal government to fund large R&D programs to find a cure. But as matters now stand, the powers that be are too willing to look the other way, even as the rising epidemic of auto deaths threatens to cause behavioral changes that can have negative implications for the American economy.

The most obvious changes show up among the growing segment of the population that is over 65. It is a known fact that people become more cautious as they grow older, more aware of dangers to their personal security, more willing to change their behavior in order to avoid perceived threats that they may have shrugged off cavalierly in their younger days. So the threat of death or serious injury from auto accidents becomes an increasing worry to a steadily aging American population.

This worry becomes more sharply focused among even broader segments of the population as local TV news shows devote increasing amounts of airtime to dramatic footage of serious traffic accidents and their resulting deaths and injuries. The significant decline in street crime across the country and the fact that the incidence of fires is at a quarter-century low have robbed local TV stations of much of their "it-leads-if-it-bleeds" news material from these sources. Hence the increasing emphasis on covering traffic accidents.

It is therefore inevitable that growing numbers of Americans, increasingly conscious of the threats to their safety that auto accidents pose, find more reasons to drive less, to shop less, to go to movies and restaurants less, and to spend more of their time in the apparent safety of their homes. And since virtually all driving trips generate at least some economic activity, the rising fear of auto accidents among large segments of the population poses the specter of less economic activity per capita in the future. Therefore, even if we were willing to take a rather chilling bean counter's view of the value of an American life, we still couldn't ignore the economic consequences of having an increasing number of these lives participating less and less in the nation's marketplaces out of fear of driving.

Fortunately, new technology can greatly minimize the deaths and injuries that occur in motor-vehicle accidents as well as reduce the incidence of accidents in the first place.

Anyone who watches weekend NASCAR stockcar races on network television is regularly treated to automobile accidents of the most spectacular kind. A group of five or ten or fifteen cars barreling along at 150 miles per hour or more in the kind of close formations we normally associate with urban expressways abruptly dissolves into chaos. Some cars ram the track's retaining walls with unbelievable force. Others spin out and are rammed broadside by cars following close behind. Still others are flung into the air like children's toys and go rolling end over end down the track, often bursting into flames when they finally come to rest.

But when it's all over, virtually every driver walks away without a scratch. This is because onboard safety technology has reached a level of effectiveness that protects drivers from death or injury even in the most horrendous imaginable multi-vehicle accidents at high speeds.

NASCAR's experience illustrates that there is, in fact, a technology solution to the issue of reducing deaths and injuries from motor-vehicle accidents in the United States. The right kind of technology can be built into automobiles to protect their passengers from harm even when their vehicles are involved in high-speed accidents on the nation's highways (where "high speeds" are less than half as great as on NASCAR racetracks).

We can even go a step further and use technology to prevent vehicles from running into each other in the first place. Some of this crash-avoidance technology is already beginning to appear on luxury automobiles (just as electric starters first appeared in the U.S. on Cadillacs in 1912). It warns drivers with flashing lights and audible alarms when they are getting too close to the car ahead or straying out of their lanes. It can even override a driver's too-slow reaction times by automatically reducing engine speed and applying the brakes.

Needless to say, critics who oppose changing the status quo (perhaps because changes remind them too painfully of how rapidly the years are passing) insist that we cannot depend on technology to ride to our rescue. This approach, they say, is riddled with practical problems that call into question its viability. For example, they argue that the cost of building all this fancy crash-avoidance and passenger-protection technology into new cars is going to further inflate their sticker prices, which are already higher than most American motorists can manage comfortably. That is why increasing numbers of motorists are turning to the second-hand market (not to mention the third-hand and fourth-hand markets) when they need to replace their cars. This is leading to a steady rise in the average age of cars on American roadways as an increasing (and now majority) proportion of the nation's private auto fleet consists of cars that are in their second decade of operation.

Canada has a lower auto fatality rate per hundred million passenger-miles than the United States. So do the United Kingdom and the rest of Europe. Ditto for Japan and Australia and many other nations that are farther down the economic development curve than the United States. We need to learn from these other nations in order to hasten the day when better onboard technology can make zero traffic fatalities a realistic goal for the United States.

NATIONAL STRATEGIC PLANNING

Academic libertarians are going to tell us that the answer to all these transportation problems lies in allowing individual entrepreneurs maximum freedom to discover profit opportunities to serve the nation's complex mobility needs through the free market. With the marketplace itself using its magic to coordinate everything so that enough of the right kind of mobility is provided in the right places.

On the other hand, devotees of top-down central planning will assure us that the answer lies in having government take a properly macro view of society's mobility needs. Working its way down through the hierarchy of details as its planners seek to determine what kind of transportation facilities are needed, in what quantities and in what places. Then proceeding to build them in accordance with a detailed Master Plan. What could be more logical?

To which libertarians will respond that planners never get it right because they can't out-think the free market.

Except, of course, when it came to winning World War II. Then the federal government unhesitatingly chose top-down, Washington-based, national strategic planning as its primary economic management tool to produce and allocate the vast quantities of fighting men, guns and ammunition, tanks, planes, and ships that overwhelmed Germany and Japan (without disrupting the civilian economy too seriously). Which most Americans understandably regard as one of their country's greatest triumphs.

When President Roosevelt delivered his ***Day of Infamy*** speech to Congress on December 8, 1941, the federal government already had a set of detailed game plans for defeating Japan in the Pacific (WAR PLAN ORANGE) and Germany in Europe (RAINBOW FIVE). These plans had been developed over the years by military planners in the various armed services and they outlined the basic strategies that the United States would follow in fighting the war.

The game plans also contained extensive detail about the numbers and types of fighting men that each of the armed services would require to carry out these strategies, how to recruit and organize them, and how to train them. Finally, the game plans detailed the types and quantities of weapons and equipment these fighting men would need, how to produce them, what kinds and quantities of raw materials their production would require, and how and where to allocate them among the various theaters of war for maximum effect.

It was all there in black and white. And as history has demonstrated, it worked supremely well.

In fact, this kind of national strategic planning worked so well that many of the managers of corporate America who had been loaned to the federal government during the war began applying its principles in their own companies after the war ended. This helped pave the way for the decades of rising prosperity that followed and came to define the core management approach in vertically integrated Fortune 500 companies.

To be clear about our terms, STRATEGIC PLANNING is the process through which we develop coherent strategies to achieve certain goals. STRATEGIC ANALYSIS AND THINKING is the part of the process where we develop strategic alternatives that leverage our strengths and minimize our weaknesses in order to exploit the opportunities and counter the threats presented by the external environment. STRATEGIC MANAGEMENT embraces the activities involved in developing our strategic plans, implementing them in the real world and revising them in light of actual experience.

Clearly, we need this kind of strategic focus at the national level if our efforts to address the nation's transportation problems are to be successful. Otherwise we risk missing worthwhile opportunities, doing new projects without proper coordination and ending up spending a lot of money to make things worse.

The Chicago Skyway Deal

For example, the Chicago city government recently leased for 99 years a 7.8-mile toll highway known as the Chicago Skyway to a foreign consortium of two private firms for an upfront cash payment of $1.82 billion. Did this action make sense? It certainly generated favorable comments from the various private organizations like Standard & Poor's that rate Chicago's outstanding debt. But their concern is limited to the city's financial position.

The city intends to use $325 million of this $1.82-billion cash inflow as a special reserve to help achieve balanced operating budgets during the next several years. It added another $500 million to its long-term reserves. Since the city has $500 million in Skyway bonds still outstanding (but has turned over to the consortium the toll-revenue stream that secured these bonds), the assumption is that it will use a portion of the $995 million remaining from the lease payment to fund Skyway bond interest and principal payments. Use of the rest of the lease payments (roughly $500 million) has apparently not yet been determined. It might be used to help fund new transportation capital projects on a pay-as-you-build basis, fund other (non-transportation) capital projects on the same basis or provide further support for the city's operating budget.

The Chicago Skyway connects the Dan Ryan Expressway in the southern part of the city with the Indiana Toll Road, which is part of I-90. It therefore provides a key link in the Interstate highway complex that runs from Chicago's Loop across northern Indiana to points as far east as Boston. The Skyway was recently restored to good condition by the city. Its tolls are relatively modest by current standards and have not been raised since 1993. Therefore, the Skyway is believed to have considerable potential for higher toll revenue that remains to be tapped.

Under the terms of the lease, the consortium is free to raise Skyway tolls within certain defined limits until 2017 and thereafter at an annual rate of 2 percent, the rate of inflation or the rate of nominal growth in per capita GDP, whichever is greater. This is expected to provide the consortium with toll rates (and therefore toll revenues) that rise significantly faster than inflation throughout much of the lease period. At the same time, the consortium is committed to investing $70 million during the next four years to improve the Skyway. This includes converting it to electronic toll collection.

On its face, this privatization deal is not unlike the one described earlier when a municipality facing a budget deficit in an election year sold off its library system to a local developer for upfront funds to close its budget gap. The most significant difference is that motorists using the Skyway (rather than Chicago taxpayers) will fund the consortium's future revenue stream. But in any case, the City of Chicago is losing control over a municipal transportation asset whose untapped toll revenue is assumed by the consortium to have a present value of at least $1.82 billion during the next 99 years.

It is easy to understand why a municipal government that is struggling to balance its budget might be tempted by such a privatization deal, since a bird in the hand is generally assumed to be worth two in the bush (though not according to the financial rocket scientists who subscribe to Real Options Theory, which can demonstrate that the opposite is often the case).

In the case of the Skyway deal, there are questions about whether the City of Chicago received fair value in the lease deal. This would be a problem if an evaluation of the alternatives shows that the true opportunity cost to the city (i.e., loss of control over a transportation asset and its future revenue stream) is higher than $1.82 billion. For example:

- **Was there meaningful competition among private-sector bidders for the lease?** The two other bids were $700 million and $500 million. But these bids appear not to have reflected the toll-revenue value during the later years of the lease. The bidders were either looking for a bargain or failed to do their homework. In any case, this scarcely provides much support for the concept of meaningful competition for the lease.

- **Could the City of Chicago have issued at least $1.82 billion in new Skyway bonds backed by future toll revenues while still retaining control of the Skyway?** Possibly not if we assume that political constraints on increasing toll rates were unavoidable so long as the Skyway was owned by the city. But there might be ways around these constraints.

- **Could the city have realized more than $1.82 billion by employing a variation of the Hong Kong model we considered earlier?** That is, the city would establish a wholly owned commercial corporation to which it would sell or lease the Skyway for an upfront payment. The corporation would fund this payment through some combination of debt (secured by toll revenues) and equity (obtained by selling minority ownership to private investors). Meanwhile, the city would retain at least policy control over the Skyway through its majority ownership of the corporation.

- **Could the city have done better financially over the long term by accepting a smaller upfront payment in return for sharing ownership with the consortium and therefore being able to share in its future financial benefits?** We know that the two consortium members recovered much of their original equity investment by replacing it with bank debt after the lease was signed. Does this suggest the possibility that they might sell the lease for a nice profit to another private investor group after Skyway tolls have been raised and the conversion to electronic toll collection has been completed? Shouldn't the city be able to share in this profit?

These questions involve only the issue of whether the City of Chicago received fair value for parting with the Skyway through a 99-year lease. They implicitly assume it is reasonable for a municipality to consider selling off certain of its physical assets for cash. In this case, "reasonable" is defined by whether the cash payment is at least as high (all things considered) as the opportunity cost of parting with the asset.

The answer could be "YES" if the resulting cash is used to invest in new capital assets whose long-term value is greater than that of the asset sold. But it would more than likely be "NO" if all or most of the cash is used to fund operating expenses. This would simply repeat the sorry pattern of the late and much-lamented ***Pan American World Airways***, which sold off or leased away in piecemeal fashion its landmark headquarters building in Manhattan, its passenger terminals at airports around the world, its fleet of airplanes and big chunks of its globe-spanning route network for infusions of cash in a vain effort to keep flying.

But the true opportunity cost of the Skyway lease deal may be considerably higher than can be measured by simple-minded financial considerations.

Suppose at some future time, for example, the Chicago metropolitan region should want to consider making its vast roadway network self-supporting from user charges on its limited-access highways. Suppose it should decide that the best way to do this is by creating an independent transportation enterprise that is jointly owned by government and private investors. What does it do about the critical highway link that is (for practical purposes) owned by the Skyway consortium? Does it simply leave out the Skyway? Is it feasible to buy out the consortium? Does the existence of the consortium create other complications that might practically rule out the possibility of a self-supporting roadway network for the Chicago metro region?

In the absence of any meaningful strategic planning for Chicago's transportation future, there is no way to know at this point whether any of these issues will ever be relevant. And that is the whole point. If we simply wander aimlessly thinking only of the present, we are bound to be surprised by what the future springs on us. And possibly kicking ourselves with regret for what turned out to be mistakes we made in the past. Like selling off transportation assets in piecemeal fashion.

Meaningful strategic planning is the only protection we have against unpleasant surprises in the future. Without it, we are simply helpless victims of Fate. But if we take strategic planning seriously and do it wisely, we have at least a chance of taking Fate by the throat and bending it to our will.

This has as much relevance for transportation today as it did for defeating Germany and Japan in World War II. That is why we need to learn as much as we can from how the federal government used strategic planning to win the war.

War Plan Orange: An Example of Strategic Management in Action

The activity of developing strategy is too often thought of as a by-the-book, one-shot undertaking to provide managers with a comprehensive playbook that is supposed to cover all eventualities. But in the real world, this is scarcely the case.

Instead, effective strategic planning is a relatively messy ***process*** for:

- Evaluating everything that we know about external reality at any given time.
- Designing strategies to accommodate reality so that we can achieve our long-term goals.
- Being alert to changes in reality through constant monitoring.
- Revising our strategies (often while we are trying to carry them out) to take such changes into account.

Military history provides one of the best sources of information about what developing strategy is all about, how to do it properly, why it must be regarded as an ongoing process and how it must respond to changing realities.

As already mentioned, the federal government had a game plan for defeating Japan in its back pocket at the time of the Pearl Harbor attack. It was known as **WAR PLAN ORANGE**, and it had been under development by the U.S. Navy since 1905. The Navy began this effort and carried it forward in response to growing awareness that U.S. acquisition of the Philippines during the Spanish-American War was likely to create conflicts with Japan in the Western Pacific that could eventually lead to war.

The development of War Plan Orange was an ongoing process that had by 1941 undergone many revisions and updates to reflect changing political and tactical realities (such as the emergence of the aircraft carrier as a naval weapons system that showed potential for becoming as important as the traditional battleship). But it was built around a common master scenario:

- War would begin with a surprise attack by Japan (probably against the Philippines). This reflected the fact that Japan had already launched such a surprise attack to begin its war with Tsarist Russia in 1904.
- Japan would immediately follow this with an invasion of the Philippines, which the United States tended to regard as impractical to defend successfully. So the U.S. Army garrison there would fight a delaying action, retreating to the Bataan Peninsula at the head of Manila Bay to deny Japan's navy use of the Bay for as long as possible.
- This delaying action would buy time for the U.S. Navy to assemble its war fleet and steam across the Central Pacific, occupying some islands along the way and bypassing others.
- The climax would come in a showdown naval battle between the United States and Japan somewhere in the waters between the Philippines and Japan, which the U.S. Navy expected to win by outnumbering Japan's warships. After which Japan would enter into negotiations with the United States to end the war and recognize American hegemony in the Western Pacific (which was always the ultimate goal of War Plan Orange).

Those of you who have read histories of the war against Japan will notice how closely War Plan Orange reflects what actually happened, even to the climactic naval battle near the Philippines.

Actually, there were two such battles within a few months of each other. The first was the ***Battle of the Philippine Sea*** in June 1944, during which the U.S. Navy largely demolished the Japanese Navy's offensive capability by destroying its carrier air forces and killing most of their pilots. The second was the ***Battle of Leyte Gulf*** in October 1944 (generally regarded as the "largest naval battle in history" in terms of the number of ships involved), when the U.S. Navy finished off most of the Japanese Navy's remaining defensive capability by sinking large numbers of its surface warships.

However, the realities of war introduced two significant STRATEGIC INFLECTION POINTS that War Plan Orange had not previously anticipated. Andrew Grove of Intel coined the term Strategic Inflection Point to describe what happens when a major change takes place in the external realities to which strategic plans must respond, even if this requires wholesale revisions. To deal with these Strategic Inflection Points in its war with Japan, the federal government had to undertake some fast-on-its-feet revisions to War Plan Orange.

The **first Strategic Inflection Point** arose when the legendary U.S. general Douglas MacArthur and his family were evacuated from the Philippines in 1942 and moved to Australia. At the time, MacArthur seriously expected to be court marshaled for royally screwing up the defense of the Philippines. This kind of disgrace had already befallen Admiral Husband Kimmel and General Walter Short, the two Pearl Harbor commanders, who were accused of "dereliction of duty" for failing to anticipate the December 7 attack and had their careers destroyed.

But MacArthur was a great favorite among Republicans in Congress, and President Roosevelt needed their cooperation in order to pursue his grand strategy with the United Kingdom and the Soviet Union for defeating both Japan and Germany. So Roosevelt chose to hail MacArthur as a war hero (the United States was badly in need of some war heroes at that low point in the Pacific War), arranged to have him awarded the Congressional Medal of Honor and placed him in command of all Allied forces in the Southwest Pacific.

These events lifted MacArthur's spirits and motivated him to develop his own game plan for defeating Japan, which had nothing to do with War Plan Orange. It involved some brilliant military hop-scotching along the eastern coast of Japanese-occupied New Guinea to neutralize its large naval base at Rabaul and end the island's military value to Japan. This would pave the way for an American invasion of the

Philippines, which MacArthur regarded as essential for public-relations reasons ("I shall return") and to serve as a staging area for an eventual invasion of Japan.

Needless to say, the U.S. Navy considered this whole idea a waste of resources that should properly be devoted to carrying out its latest version of War Plan Orange. In fact, it so outraged Chief of Naval Operations Admiral Ernest King that it led to his daughter's legendary observation that her father was the most even-tempered man in the U.S. Navy—he was always in a towering rage.

Fortunately, Roosevelt's earlier strategic concept of turning the United States into the "arsenal of democracy" was finally bearing fruit in wholesale quantities as increasing portions of the nation's productive capacity became devoted to turning out the fighting men and weapons needed to wage a two-front war. This enabled Roosevelt to accommodate the new realities of this Strategic Inflection Point and win further brownie points among Congressional Republicans by providing MacArthur with the military resources he needed without shortchanging the Navy.

Thus the United States was able to carry out *two* simultaneous game plans for defeating Japan—MacArthur's Southwest Pacific strategy (which was driven by political realities in Washington) and the Navy's War Plan Orange strategy in the Central Pacific (which most military historians regard as the one that really mattered).

The **second major Strategic Inflection Point** affecting War Plan Orange was the collapse of its implicit assumption that Japan would begin peace negotiations once its navy had been defeated, with no further military action being needed.

By January 1945, it was clear that Japan had lost the war. It was cut off from all its Asian conquests due to the loss of its navy as an effective fighting force. It could no longer import the food, petroleum and other essentials on which it depended to continue functioning as anything like a normal society because its merchant fleet had been sunk by U.S. submarines. And it was now within range of U.S. heavy bombers as well as U.S. carrier air forces and shore-bombardment battleships. Its only rational course was to try to cut the best deal it could with the United States through peace negotiations.

But Japan's government was dominated by Army generals who, in the best Samurai tradition, insisted that national suicide in the name of the Emperor was a better alternative than the humiliation of a Western-style negotiated peace. This Strategic Inflection Point required the U.S.

to quickly develop some major new amendments to War Plan Orange to guide its end game against Japan—amendments that could impose severe enough pressure on Japan's government of military fanatics for it to be replaced by a government of civilian peaceniks who would be willing to negotiate. These new amendments would have to involve the right balance among four options:

- **Imposing an aggressive sea and air blockade against Japan to literally starve it into submission.**
- **Mounting a land invasion to occupy Japan by armed force (as was necessary to bring about Germany's surrender).**
- **Launching and sustaining an all-out aerial bombing campaign to destroy Japan's cities and internal transportation links.**
- **Encouraging the Soviet Union to enter the war against Japan as soon as possible in order to take advantage of the enormous military power of the Red Army.**

Given the realities of the situation, there was no way to determine in advance which option would be most effective (though each one had its own group of eloquent advocates in the federal government). The United States would have to play it by ear. This meant moving ahead with all four options simultaneously, seeing which ones seemed to be working best, and modifying end-game strategy on the fly to exploit the opportunities presented by the external environment.

By itself, the BLOCKADE OPTION would be the least costly for the United States since it would involve few American casualties. But the experience of Leningrad during World War II raised some troubling questions. Leningrad had withstood a three-year siege by Germany that cost the city more than a million dead civilians before the Wehrmacht finally gave up and withdrew. So it seemed likely that the blockade option against Japan could take several years to bear results. And the federal government was concerned about whether the U.S. population (which is not noted for having a very long attention span—witness Vietnam and what now seems to be happening with Iraq) would have enough patience to wait long enough for Japan to cave, since it was already weary from having endured three and a half years of war against Germany and wanted to "bring the boys home."

At the other end of the spectrum was the INVASION OPTION, which promised maximum casualties for both sides (up to half a million dead and wounded among U.S. ground forces and "untold millions" among Japan's army and civilians).

This option was essentially a re-run of the earlier invasion of Germany. Since the Nazi government was as irrational as Japan's and insisted on fighting to the bitter end, the occupation of Germany by armed force seemed like the only feasible way to end the European war. So the U.S. and U.K. armies (invading from the west) and the Soviet Red Army (invading from the east) fought their way into Germany during the early months of 1945 and laid waste to the country.

But the heavy casualties suffered by the United States during the ***Battle of the Bulge*** (Germany's last-gasp counter-offensive that ended in January 1945) made both the United States and the United Kingdom especially sensitive to the public-relations problems of incurring large numbers of dead and wounded among their armies. So their implicit policy became one of encouraging the Red Army to do most of the fighting (and therefore most of the dying) during the invasion of Germany. The price of this was to turn the eastern half of Europe over to Soviet domination, which became the basis for the Cold War.

Given a possible invasion of Japan, should the United States assure itself of sole control over the postwar occupation of Japan by doing most of the ground fighting itself and absorbing most of the Allied casualties? Would the war-weary American population sit still for this? Or should the United States try to minimize its casualties by getting the Red Army to participate (see below), and accept some sort of partnership with the Soviets for the occupation of Japan? As it turned out, these questions became moot when a new Japanese civilian government surrendered in August 1945, several months before the land invasion was scheduled to begin.

In many respects, the AERIAL BOMBING OPTION was the most tantalizing, at least on paper.

In the wake of the First World War, visionary military officers in the United Kingdom, Italy and the United States began speculating about the possibility that ***Strategic Aerial Bombing*** could win future wars all by itself. After all, the theory went, if you exploited rapidly developing aviation technology to destroy enough of a nation's arms-producing factories and terrorize its urban populations with aerial bombing, how could that nation not surrender fairly quickly without any involvement by ground or naval forces? This simple logic seemed unassailable to large portions of the Western world's civilian populations and their political leaders—especially in view of the devastating results from Germany's bombing of Guernica in Spain, Warsaw in Poland and Rotterdam in the Netherlands, plus Japan's bombing of Nanking in China.

Once the United Kingdom and the United States became involved in World War II in Europe, there emerged two distinctly different tactical approaches to implementing the aerial bombing strategy.

The **U.K. APPROACH** involved nighttime fire raids to burn down German cities. RAF leaders argued that most of Germany's military factories, transportation junctions and electric utility plants (not to mention the workers who operated them) were located in cities. So burning down cities would cripple Germany's war-making capability and force it to surrender without any need for a high-casualty land invasion of Germany (the prospect of which is reported to have given Winston Churchill such nightmares that it led to year-by-year postponements between 1942 and 1944 of the amphibious landing in France). Also, a city happened to be the smallest target that the RAF could find and hit at night—though that was something RAF leaders chose not to talk about.

The very different **U.S. APPROACH** involved something it called ***Daylight Precision Bombing***. Leaders of the U.S. air forces were committed to the idea that the super-sophisticated technology of their Norden bombsight made it possible for B-17 and B-24 heavy bombers to destroy individual military targets (fighter aircraft factories, oil refineries, ball bearing plants, etc.) by pinpoint daylight bombing without causing collateral damage to surrounding urban residential areas. This was expected to cripple Germany's war-making capability quickly, efficiently, and with minimum civilian casualties, thereby bringing a rapid end to the European war. (And, not so incidentally, bringing to fruition the long-held dream of the air generals for turning what was then the U.S. Army Air Forces into an independent service enjoying equal status with the Army and the Navy.)

But even after the U.S. air forces gained control of the skies over Germany by destroying the Luftwaffe's fighter arm (in early 1944) so that its bombers had little opposition, there were still problems in making Daylight Precision Bombing work. Since the Norden bombsight was optically based, its use required that bombardiers be able to see their targets in order to be able to bomb with precision. But weather conditions in Germany meant that targets were frequently obscured by cloud cover. Under such conditions, the U.S. air forces had no choice but to bomb using radar. This technology was so crude in those days that it turned the hallowed concept of precision bombing into RAF-style area bombing.

In any case, the bombing of Germany by the U.S. air forces and the RAF ultimately caused immense damage to urban areas and severely disrupted German life. Air-force leaders claimed that bombing shortened the European war by at least a year. Maybe so. But one thing is clear: bombing couldn't eliminate the need for a land invasion to defeat Germany.

However, U.S. air-force leaders were convinced that things would be different in Japan. For one thing, the technological innovations built into the new B-29 now gave them a bomber that could fly farther, higher, faster and carry a heavier bomb load than the bombers they had used against Germany (the B-29 program was the most expensive U.S. weapons system ever undertaken to that point—half again as costly as the development of the atomic bomb). For another, Japan's aerial defenses against bombers were believed to be much cruder and less effective than Germany's. So air-force leaders insisted that strategic bombing could force a Japanese surrender without any need for a costly land invasion (and thereby enable them to achieve their ambition of an air force that was an independent service rather than a stepchild of the Army). The U.S. air forces began their Daylight Precision Bombing campaign against Japan late in 1944—initially from bases in northern China, then from much closer bases in the Mariana Islands.

But again there were problems. One was frequent cloud cover over target areas (just as in Germany) that interfered with precision bombing. Another was that the high-altitude flight paths of the B-29s exposed them to the effects of the then-unknown jet stream, whose high-speed winds distorted their effective ground speeds in ways that the target-tracking analog computers built into the Norden bombsight couldn't handle. So bombing accuracy was very poor and air-force leaders had to face the fact that their long-cherished Daylight Precision Bombing concept wasn't working.

Very reluctantly and after much soul searching (accompanied by some inevitable top-level personnel changes), air-force leaders decided to try the RAF concept of nighttime fire raids against Japan's cities. The first of these raids took place against Tokyo on the night of March 9, 1945—with results that were more spectacular than anyone had anticipated. Some 282 B-29s dropping 2,000 tons of napalm bombs from altitudes of only 7,000 feet generated a massive firestorm that turned 16 square miles of northeast Tokyo into a burned-out graveyard filled with the charred-beyond-recognition remains of at least 100,000 Japanese civilians.

This was by far the most devastating single air raid in the history of human warfare. It greatly exceeded the death and destruction from the later atomic-bomb raids against the much smaller cities of Hiroshima and Nagasaki (though these raids received much more international publicity and have assumed something like mystical significance). And its success launched the U.S. air forces on a massive fire-bombing campaign that burned down 66 Japanese cities, turned most of the vast Tokyo/Kawasaki/Yokohama metropolitan region into an uninhabited moonscape and culminated in the August atomic-bomb raids against Hiroshima and Nagasaki. In fact, the air forces had to cut back their bombing campaign in mid-summer 1945 because they had simply run out of targets (except for the cities that had been reserved for the atomic bomb).

The SOVIET UNION OPTION was clearly a wildcard and one that the federal government debated at length. It was also the option over which the U.S. had the least control.

Stalin had promised Roosevelt and Churchill that the Soviet Union would enter the war against Japan soon after Germany was defeated, although his timetable and price for doing so remained unclear. The military advantages of having the Red Army involved in the fighting were obvious to everyone, because it was probably the most successful army in history (certainly when measured by the thousands of square miles of territory it had liberated or conquered in the war against Germany). There was also the psychological impact on Japan's government of being confronted with the Red Army's enormous reserves of fighting men, tanks and field artillery. Not to mention the fearsome brutality of its all-out approach to making war (which, as history has demonstrated, burned away forever the German militaristic traditions that had periodically disrupted the peace of Europe since the middle of the 19th century). There was little question that these advantages could greatly shorten the war.

On the other hand, the United States couldn't help being concerned about the potential complications of having to share the postwar occupation of Japan with the Soviet Union. But Washington could do little about this except to hope that the other options would cause Japan to surrender before the Soviet Union became involved.

As it turned out, the Soviet Union didn't go to war against Japan until two days after the Hiroshima bomb was dropped on August 6 (and one day before the Nagasaki bomb was dropped). The Red Army quickly overran Manchuria and occupied the southern half of Sakhalin Island (the northernmost Japanese home island).

But on August 14, after surveying the destruction in the Imperial Palace compound in Central Tokyo, the Japanese emperor broke with long tradition by personally intervening to order Japan's government to seek peace negotiations with the United States. He then did the unimaginable by going on the radio to tell the people of Japan that the war was over and they must henceforth "bear the unbearable and endure the unendurable." And Stalin made no serious attempt to insist on sharing in Japan's occupation.

In summary, the United States proceeded to move forward with all four options during 1945. But it actually implemented only the blockade and aerial bombing options. Japan's surrender made the invasion option unnecessary, while the Soviet Union option came too late in the day for U.S. fears about its postwar complications to be realized.

Historians have long debated which option (or combination) brought about Japan's surrender. The popular view is that the spectacular effects of the two atomic-bomb raids turned the tide. Since they were logical extensions of the aerial bombing option, this would provide the only real-world vindication of the theory that strategic bombing could win wars all by itself. For various reasons, it couldn't do this in Germany, Korea, Vietnam or in either of the two wars against Iraq. But the Japanese example did pave the way for the U.S. Army Air Forces to finally become an independent armed service and made the threat of strategic bombing (with nuclear weapons) a core U.S. strategy during the Cold War.

However, some historians believe that Soviet entry into the war was the main factor in causing Japan to surrender. Others believe that, by late summer, the effectiveness of the blockade option was confronting Japan with mass starvation in a matter of a few months. Presumably, this might have forced the Japanese government to see reason without any need for the United States to invade Japan, drop atomic bombs or continue other devastating forms of strategic bombing that had already turned most of urban Japan into wasteland. (But who knew how long the Soviets would wait to enter the war?)

History provides an interesting context for evaluating these arguments. It involves the fact that the Japanese army followed a practice of engaging in the wholesale slaughter of civilians in the Asian countries that it occupied. This led to the deaths of more than 15 million people in China alone (nearly three times as many civilian deaths as in the European holocaust). And it was continuing without pause during 1945. Therefore, the need to stop this slaughter called for using any means available to induce Japan to surrender as quickly as possible.

It is doubtful that the U.S. government was aware of the extent of this ongoing slaughter, though the liberation of the Philippines by the end of 1944 provided dramatic evidence of the Japanese army's treatment of Asian civilians.

But given the Japanese government's unwillingness to begin serious peace negotiations before late summer 1945, it is not unclear whether the U.S. government could have done anything else to speed up the end of the war. So whether we attribute Japan's final surrender to the two atomic-bomb raids, the devastation of Japanese cities by conventional fire bombing, the effectiveness of the blockade in bringing Japan's population to the verge of starvation or Soviet entry into the war, at least the slaughter of civilians throughout occupied Asia by the Japanese army finally came to an end by the autumn of 1945.

The moral overtones of this reality may outweigh the various criticisms of U.S. wartime conduct against Japan that have become fashionable in revisionist circles down through the years.

The Lessons of War Plan Orange

War Plan Orange is one of history's best examples of successful strategic management in action. So it has much to teach us about how to use the same thing to guide us in solving U.S. transportation problems.

The **first lesson** is that strategy must reflect reality, not just our hopes about the way we think reality ought to be. This requires lots of hard information about the current nature of reality, and lots of analysis to determine what this information is actually telling us.

Until 1941, the development of War Plan Orange was something of an abstract exercise because the United States was not at war with Japan. But its planners sought to reflect the latest relevant information in their numerous revisions. This information came from the results of war-game exercises that explored new approaches to naval tactics (such as the use of aircraft carriers as offensive weapons that could rival battleships), new wrinkles in warship design that extended their war-making capability (such as the use of radar to direct naval gunfire), and political intelligence about evolving Japanese ambitions and capabilities (such as its invasions of China and French Indo-China, and the ramping-up of its naval construction program during the 1930s).

In transportation, much of the information we have is antiquated in terms of what it measures. We have lots of reasonably current statistics about engineering-oriented variables like trip volumes—the number of annual passenger-miles on highways, ton-miles of goods moved by rail, and so on. But we lack meaningful information that is focused on the all-important reality of SERVICE TO CUSTOMERS. This includes such critical measures as average door-to-door travel times for people and goods, how widely these averages vary (which is a measure of transportation reliability), and how much travelers are spending per mile to move people and goods, both on a simple out-of-pocket basis and in terms of how much time its takes to get from here to there.

Clearly, our approach to collecting information about transportation needs a major overhaul of its content and scope in order to make it more relevant to the task at hand.

The **second lesson** is to recognize that strategic management must be an ongoing process in order to accommodate changes in the realities of the external environment. The world is constantly changing, and each change has the potential for uncovering new Strategic Inflection Points that define the reality of the changes to which managers must respond.

In the case of War Plan Orange, we saw how the United States had to quickly develop some brand-new plan amendments in early 1945 to deal with the fact that Japan's government refused to begin peace negotiations after the defeat of its navy in the Western Pacific (as the standard plan had always expected). The U.S. response to this Strategic Inflection Point resulted in the "four options" end-game strategy and the need to move ahead with all four options more or less simultaneously because there was no way to determine in advance which one would be the most effective.

In the transportation arena, a major Strategic Inflection Point is the emergence of new technology and the many opportunities it offers. As we've seen, new technology now allows us to price highway use by distance traveled and other key variables. This provides us with an opportunity to revolutionize the way that we fund and manage roadway networks and, in the process, create a new revenue source for the entire spectrum of surface transportation in the United States. This Strategic Inflection Point makes obsolete existing assumptions about how, by whom and to what extent improvements to roadways and other transportation modes can be financed. New directions in strategic planning for the nation's transportation systems are needed to exploit the opportunities that new technology offers.

The **third lesson** to be learned is that strategy development and implementation go hand-in-hand. It may be a nice classroom convenience to talk about strategic planning as if it can be a form of immaculate conception that ignores the unfashionable business of implementation. But the real world presents us with a far messier environment. Without the necessary feedback from implementation, planning becomes a static playbook for winning a theoretical football game that usually turns out not to materialize on game day. So planning must always be ***entrepreneurial.*** It must always be looking for new opportunities in the real world to exploit strengths and avoid weaknesses.

Until late 1941, War Plan Orange had always assumed that Japan would begin war with the United States with a surprise attack against the Philippines or somewhere else in the Western Pacific. Instead, the surprise attack was against Pearl Harbor. Its immediate effect was to cripple the U.S. Navy's battleship fleet, thereby robbing it of what had always been regarded as its principal offensive force. While the United States had a massive program underway to construct a fleet of new battleships (including those of the superlative Iowa class), it would not begin to bear meaningful fruit until late 1942. This Strategic Inflection Point forced a complete rethinking of how to mount a rapid offensive response against Japan.

Fortunately, the Navy's "brown-shoe" aviation admirals had been struggling for years against tradition-oriented "black-shoe" battleship admirals to develop aircraft-carrier aviation as an offensive force in its own right. So with no other choice, the United States turned its brown-shoe admirals loose to test the actual offensive power of task forces built around aircraft carriers.

A series of hit-and-run carrier raids against Japanese island bases in the Central Pacific during the early months of 1942 demonstrated that such taskforces could become a new U.S. offensive strength to offset the unanticipated weaknesses arising from the loss of battleships at Pearl Harbor. Equally important, the real-world experience gained from these raids enabled brown-shoe admirals to refine their implementation tactics and expand their strategic understanding of the broader role that carriers could play in implementing War Plan Orange.

The culmination of these early efforts came on June 4, 1944 at the ***Battle of Midway***—when a combination of breaking the Japanese naval code and the ability to exploit opportunities presented by Lady Luck herself enabled 50 dive bombers from the U.S. carriers ***Enterprise*** and ***Yorktown*** to catch Japan's carrier fleet with its pants down. Diving out of the sun like Destiny's own arrows, these bombers demolished three of Japan's four first-line aircraft carriers in a mere five minutes, broke the back of Japanese naval aviation power and changed the course of the Pacific war.

It is worth noting that the tactically brilliant and daring Japanese attack on Pearl Harbor actually turned out to be a major strategic error. In planning the attack, Japan's navy was unable to look beyond the end of its nose and therefore focused its efforts on crippling the U.S. Navy's fleet-in-being, which turned out to consist of obsolete battleships whose day was past.

A more effective choice would have been to focus on crippling Pearl Harbor itself as a mid-Pacific naval base by destroying its fuel storage and ship repair facilities. This would have driven the U.S. fleet back to the West Coast of the United States and prolonged the Pacific war by several years.

The U.S. transportation world is unlikely to experience anything as spectacular as the Pearl Harbor and Midway dramas. But the principles that they represent are certainly applicable. These principles involve realizing that management-by-crisis is more likely to be the norm than the exception in the real world; that critical information from experience with implementation must be fed back rapidly to jolt planners into developing new and more realistic strategic alternatives; that no strategic plan survives unbloodied its first confrontation with reality when implementation begins; and that we must always, always approach strategic management with one foot firmly planted on the shifting sands of the real world.

For example, the initial response of many tradition-oriented toll-road managers (the "black-shoe" transportation admirals) to the appearance of electronic toll-collection technology was that it would enable them to reduce their labor costs by substituting technology for people. But this narrow cost-accounting focus missed the whole point.

We are now beginning to realize that this technology's main benefit is to provide opportunities for a new strategic focus on serving customers by changing the way we finance and manage roadways; by creating new revenue sources that can pave the way for integrating roadway networks more effectively with other travel modes; and by better exploiting the latent strengths of transportation to help fuel economic growth. So the main purpose of strategic management is to guide a process for recognizing such opportunities, shape responses to exploit them and refine these responses through feedback from implementation in as close to a real-time environment as is possible.

Who Should Do the Nation's Transportation Strategic Planning?

The obvious choice is to assign this task to the federal government.

But the federal government of today is nothing like the competent federal government of World War II. It appears to lack any sense of where the nation's real priorities should lie, and its management fumblings in Iraq and in responding to Hurricane Katrina seem like something out of a classic Laurel and Hardy movie (although more tragic than funny). For some observers, the prospect of turning the important responsibility of developing a master game plan for improving the nation's transportation system over to the federal bureaucracy scarcely inspires much confidence.

Nearly 75 years of New Deal activism, World War II triumphs, Cold War chess games and post-9/11 hand wringing have given us a knee-jerk sense that national problems are automatically the turf of the federal government. The feds certainly appear to believe this, which may be why they fail to recognize the possibility of any competition in this area. And with no awareness of the potential for competition, it is easy for the federal government to comfortably assume that we will (like Oliver Twist's fellow orphans) humbly take whatever it gives us and never ask for "more."

But suppose we were to stand up on our hind legs and insist on "more"—even to the extent of looking in other corners of society for institutions that seem capable of delivering the goods more effectively and responsively than the feds. Suppose the feds should wake up one day and hear us say: Sorry, but we don't need you guys for this particular problem. We've found other options that look like they'll work better.

Suppose this should happen enough times on enough problems for the federal government to seem increasingly irrelevant. Could this lead to a growing chorus of questions about whether it makes any sense for the federal government to continue grabbing the lion's share of the nation's total tax dollars? And if this chorus becomes loud enough, might it not result in a national consensus to strip away much of the federal government's present taxing power so that more dollars would be available for American society to fund its new problem-solving institutions? What role would be left for a revenue-starved federal government in such a case?

Is this simply idle speculation? Maybe not. Because there is evidence that the natives are growing increasingly restless.

For example, the State of Louisiana is surely not on anyone's top ten list for having effective government. But in the wake of the federal government's inept response to Hurricane Katrina, Louisiana has proposed that a nine-member independent commission be established to oversee the planning and management of a series of Army Corps of Engineers projects to protect its coastline and New Orleans from Category 5 hurricanes. The Louisiana government would appoint at least five members of the commission in order to assure local control. The White House and Congress would delegate to the commission the power to authorize projects, allocate construction funds and supervise the work of the Corps of Engineers on these projects. The federal government's role would be limited to writing checks to cover projects costs, which are anticipated to total some $40 billion.

This proposal builds on the recent success of the quasi-independent commission to develop a list of military bases that should be closed or combined in the interests of economic efficiency. The political implications of developing such a list made this a task that was clearly beyond the capabilities of Congress or the White House. But assigning the task to an independent commission essentially flipped the issue of responsibility on its head. The existence of such a third-party list meant that Congress or the White House now had to make a case for removing any particular base from the list. So the political heat involved in not removing most bases from the list was regarded as considerably more temperate.

In this context, some visionaries have suggested establishing an independent NATIONAL TRANSPORTATION COMMISSION to assume responsibility for planning and managing the restoration of the nation's transportation systems. The attractiveness of this proposal even led to Congress and the White House agreeing to add to the recently approved Safe, Accountable, Flexible and Efficient Transportation Equity Act (SAFETEA) provisions to create two separate commissions to study transportation issues.

Other visionaries have more grandiloquent ideas. They see the nation's transportation responsibilities as being driven by a new emphasis on economic development for domestic and international trade purposes. For them, technology has the potential to break down existing institutional barriers and speed the evolution of a more horizontal and decentralized transportation environment that is radically different from today's ossified environment. For example:

- ***The U.S. Department of Transportation*** would become more streamlined and integrated. Instead of having separate administrations for highways and transit, it would have a single SURFACE TRANSPORTATION ADMINISTRATION responsible for overall standards, safety, security and accountability issues. The new STA would also oversee replacement of the worn-out Interstate Highway System with a national network of ***integrated transportation corridors*** that would include general lanes for motor vehicles, dedicated lanes for trucks, and rail lines for passenger and goods movement. Construction and management of many of these corridors would be franchised to consortiums of private firms. They would raise capital from large new private-sector transportation infrastructure investment funds on the basis of their revenues from motor-vehicle tolls and mileage fees charged to the private companies operating trains on corridor rail lines. This reliance on user charges would enable the federal fuel tax to be largely phased out.
- ***State Departments of Transportation*** would follow the same general principles. They would act primarily as administrators of contracts awarded to private firms for transportation facility design, construction, operation and maintenance. But they would still retain responsibility for planning, budgeting and meeting federal standards. The states would have the power to levy sales taxes on motor-vehicle fuel and authorize the use of motor-vehicle tolls on non-federal portions of their roadway networks. The later would make it possible for metropolitan regions to convert their roadway networks into self-supporting enterprises owned by a mixture of local-government and private-investor partners.

Many of these aggressive concepts may seem distinctly ***Jeffersonian*** to those who have read the standard books about post-colonial American history. But this misses an important point. We should remember that Thomas Jefferson was born rich as the pampered son of a plantation owner and depended on slaves to produce a comfortable income from the land that he inherited. So he never had to confront the ice-water-in-your-face reality of having to earn his own living. This left him free to spin his bucolic visions of a rural American fairyland full of independent farmers, where public-sector initiatives were largely left to the states rather than to a strong national government.

Actually, the concepts described above draw more on the philosophical heritage of Alexander Hamilton (whom Jefferson detested). In classic American fashion, Hamilton was born poor and had to claw his way up the financial and political ladders on the strength of his always-pragmatic intelligence and his ability to see workable connections between vision and reality. That is why, for example, he was willing to accept relocation of the nation's political capital to the Potomac River marshland in return for allowing New York City to become the nation's commercial capital, and therefore the primary source of financing for its political life. And why, in a more contemporary context, it makes sense to give Congress total control over the allocation of revenues from the residual federal fuel tax—which would extend to funding pork-barrel projects—in return for its cooperation in getting other transportation legislation passed.

The key here is pragmatism. The concept of developing new institutions to take over strategic-planning responsibilities that the federal government can't or won't handle is very much in the spirit of Hamilton's pragmatic approach to solving problems. Ditto the concept of making roadway networks self-supporting through highway user charges, and allowing Congress to allocate revenues from a residual federal fuel tax as it wishes. Such concepts respond to the evolving realities of 21st-century America, which have nothing to do with the never-was America of Thomas Jefferson's post-colonial visions.

After all, Jefferson was never able to bring himself to free from slavery his companion Sally Hemmings and the children they had together, despite their obvious importance to him. So even when it came to dealing with one of the most compelling set of circumstances in his personal life, he remained a prisoner of his cotton-wool dreams.

Transportation Management

We have referred several times to the importance of managing the nation's transportation systems more effectively. But let us be clear about the kind of management style we're talking about.

It isn't like that of the typical high-school football coach, who calls every offensive play and defensive formation himself and whose players on the field simply follow orders as best they can (even when the other team does something unexpected).

Instead, we're talking about a management style that's closer to that of an NBA basketball coach. Players on the court have to respond instantaneously and like a well-integrated chamber music quintet to the challenges and opportunities presented by the other team, and these can rarely be anticipated with much certainly. Therefore, the coach has created in his players an instinctive awareness of the particular game environment (which is always changing) and the importance of being alert to unexpected opportunities.

To a savvy transportation manager, this means knowing in advance what you are trying to accomplish, determining how existing travel modes and new technology can help you accomplish this, and looking for opportunities to use the various aspects of transportation in ways not previously contemplated.

A real-world example illustrates what this means.

As noted above, when electronic toll-collection technology appeared in the early 1990s, the managers of some older, tradition-directed toll authorities saw it as a way to help them keep their unionized labor costs under control. In their calculation, it could replace the manual tasks involved in collecting tolls the old-fashioned way. So managing this technology had a narrow cost-accounting focus that they measured by "revenue per employee."

The managers of other toll authorities recognized that this technology could collect tolls anywhere on the roadway without interfering with vehicle speeds. Therefore, it eliminated the need for toll booths, land-hungry toll plazas and long lines of motorists queuing up to pay tolls. If they were expanding their scope by building new toll roads, they could use this technology to make their operations smoother and more efficient. So their management of this technology had a strong operations focus measured by "vehicle throughput per lane-mile."

Still other toll road managers saw things in a very different light. They realized that the ability to collect tolls anywhere on the road while they monitored where each vehicle entered and left meant they could offer motorists more travel choices than ever before. Such choices could involve how much of a particular trip to make using the toll road (because they could charge motorists so much per mile traveled rather than simply a flat rate), when to make a trip (because they could charge motorists higher rates during peak commuting periods and lower rates at other times), and what kind of vehicles to use for a trip (because they could charge heavy trucks a higher rate than compact sedans). All of these changes turn toll-paying motorists into ***customers making choices.*** So this kind of management had an obvious customer-choice focus, which could be measured by "average customer-miles of travel per hour" and by the variations of this average.

The next step (hopefully) will bring us an expanding group of transportation managers who understand how smart cards, more sophisticated transponders, built-in location-detection systems and the other grandchildren of the original electronic toll-collection technology can be used to establish a new source of funding for all surface transportation modes—exploiting the principles of customer choice to generate new dollars.

To these visionaries, management becomes nothing less than a basic tool for managing entire "integrated portfolios" of surface-transportation modes, just as is done in the most sophisticated multi-product corporations. This might begin at the level of metropolitan regions, eventually moving up the hierarchy to entire states, then to clusters of closely linked states, then finally to broad regions of the nation. The result would be a virtually seamless and reliable transportation network that can make the time dimension increasingly irrelevant.

The availability of new technology may be a necessary condition for integrating various surface-transportation modes into the kind of smoothly functioning and highly reliable intermodal travel continuum that the 21st century demands. But it is scarcely a sufficient condition. Equally important is a new understanding among both transportation providers and consumers. If infrastructure and its managers are to connect with vehicles and their operators in a meaningful way, both parties must come to recognize the many different levels of integration the new technology makes possible.

This must reflect the critical links between the drivers of vehicles and the companies that make these vehicles, between the shippers and receivers of freight and those who move freight, and between private enterprises seeking higher profits and public enterprises seeking the most effective use of government dollars in supporting the national economy. It must integrate paying for the use of transportation infrastructure, managing this infrastructure and marketing its use, developing the new technology it needs to improve service to customers, and designing the financing systems that expedite the smooth and logical flow of funds among all these components.

This is the kind of bold new vision for surface transportation that the United States needs if it is to truly prosper in the 21st century rather than simply limp along trying to play catch-up. It is the long overdue replacement for the last-century vision embodied in the Interstate

Highway System, which used imagination and concrete to remake an America that still thought of itself in horse-and-buggy terms.

The new vision we need must employ imagination and information to connect people and organizations, expand mobility, improve safety, raise productivity for moving both people and goods, and upgrade the management and output of the physical transportation system. This means learning how to use technology to manage the evolution of diverse surface transportation modes in the direction of becoming a truly unified system.

From one perspective, the history of technology is on our side because it suggests that the basic physical tools are coming, since technology has always had an inevitable upward curve. What matters is how we manage its use. It may be too early to say whether integrating technology into the management of the surface-transportation system can materially reduce the need to build more capacity. But it's obviously going to make more of a difference on this score than continuing to do things the old-fashioned way.

In the end, the most critical measures of how well our transportation systems are serving us may be those for which no numbers are currently being collected on a comprehensive basis. But one of these measures is likely to be average trip time per mile of travel. The smaller this measure (and the extent of its variations) becomes, the better it is for the American economy.

Transportation in 2030?

During the summers of 1939 and 1940, as war clouds broke over Europe, the legendary New York World's Fair offered an intriguing picture of what "the World of Tomorrow" could be like. One of its most popular exhibits was the General Motors Futurama, which was devoted to the promise of transportation in a paradise world of arrow-straight limited-access highways with automobiles speeding along in a congestion-free environment under automatic control.

After more than 60 years, much of this transportation paradise has yet to materialize. But it could finally be edging into the realm of the possible. As transportation increasingly becomes a make-or-break element in the nation's struggle to keep its economy growing, we will have no choice but to marshal the resources of new technology, planning and management in more intelligent and innovative ways. So another look at "the World of Tomorrow" 2030-style may be worthwhile.

At six o'clock on a September morning in 2030, an investment banker climbs out of bed in her apartment in the Art Deco elegance of the San Remo on Manhattan's Central Park West. Just like her mother some 25 years earlier, she is scheduled to fly to Cleveland for an important meeting to settle the final details of her client's upcoming common-stock offering. But New York City is experiencing one of its rare hurricanes and she worries about what this may have done to flight schedules at LaGuardia Airport.

Over a steaming cup of coffee, she turns on her personal computer and surfs to her Mobility Manager's web site where her ID number, planned door-to-door travel itinerary to Cleveland and destination arrival-time requirements are on file. She clicks on the "update" button and a second later gets the bad news she feared. All flights out of LaGuardia have been canceled until further notice because of the hurricane.

Fortunately, the "travel alternatives" button is already flashing green on her computer screen. She clicks on it and the screen quickly changes to her proposed "best alternative" travel itinerary complete with times, which appears as a series of bullets.

- She will take the Eighth Avenue subway line from the station on her corner to Penn Station.
- At Penn Station, she will board the next high-speed intercity train for a two-and-a-half hour trip to South Station in Boston (which the hurricane is not expected to reach until late afternoon).
- From South Station, she will take an express bus to Logan Airport where all flights will still be operating normally.
- At Logan Airport, she will board a flight for Cleveland's Hopkins Airport, where a rental car and driver will meet her for the half-hour drive to her client's offices in suburban Bedford Heights.

She scans the bullets, then clicks the "approve" button on the screen. A second later, the screen changes again. This time it shows the bullets with the word Confirmed next to each one. She prints out a copy for herself and forwards copies to her secretary and to her client. She also attaches a note to the copy forwarded to her client asking that their meeting be pushed back two hours, indicating that she will confirm this in a few hours by phone. Then she logs off and prepares to get dressed, deciding that she'll wait until after the Cleveland meeting is over to contact her Mobility Manager about the fastest way to get home.

Remember that family-owned chain of apparel stores on Long Island? By 2030, it has doubled in size and is now managed by the youngest son of the now-retired manager whose struggles to keep on top of his customers' buying desires we considered earlier (for the record, his eldest son is a vascular surgeon in Brooklyn while his middle son is assistant curator of an art museum in Texas).

The young apparel chain manager is even more market-oriented than his father was. In fact, he knows that success no longer depends merely on keeping up with customer buying desires. He must now try to anticipate these desires. That is why he has installed interactive television screens in all his stores that display various proposed styles of slacks, blouses and other items of female apparel for his mostly teenage customers to rate (on a five-step scale from "Must Buy" to "Wouldn't Be Caught Dead Wearing").

In his home office on a Sunday evening two weeks before the peak of the Back-To-School buying rush, the manager reviews the latest apparel-style ratings on his personal computer along with the weekend's item-by-item sales volumes. After several hours of detailed spreadsheet analysis, he has compiled a list of the new-style items to order from his supplier in a small town near Mexico City, broken down by the items for each of his stores and their volume. He emails this list to his supplier, who is waiting for it. An hour later, he and his supplier are on the phone together to discuss the details. They agree that the apparel items will be ready for shipment late Wednesday afternoon, packed into containers and sorted by store.

Then the apparel-chain manager logs on to his Mobility Manager's web site and emails the shipment requirements for this apparel order. In this case, the Mobility Manager is a third-party shipper (like UPS) who will handle all aspects of transporting the apparel containers from the supplier's factory outside Mexico City to each store on Long Island. Within an hour, the shipper emails the manager the shipment's itinerary.

- One of the shipper's trucks will pick up the apparel containers from the loading dock of the supplier's factory late Wednesday afternoon and drive them to the shipper's air-cargo terminal at the Mexico City airport.
- The shipper will load the containers onto one of its all-cargo airplanes for the direct flight Wednesday night to the shipper's hub terminal at the Louisville airport in Kentucky.

- Very early Thursday morning, the shipper will transfer the apparel containers to a smaller cargo airplane flying directly to its cargo terminal at MacArthur Airport on Long Island later that morning.
- Early Thursday afternoon at MacArthur Airport, the shipper will transfer the store-tagged containers to its local trucks for delivery by 5 p.m. to the apparel chain's individual stores.
- Each store's stock clerks will unpack the containers Thursday evening and place the various apparel items on the selling floor racks, in plenty of time for the expected rush of Back-To-School customers beginning Friday afternoon.

After exchanging emails, the apparel-chain manager and the shipper agree on price and other details for the shipment. Finally, the exhausted manager logs off and goes to bed, confident that his stores will be ready for the coming weekend's buying spree. It is now almost one-thirty Monday morning.

These two illustrations of how transportation must evolve in the future reveal three critical points.

First: Travel customers will increasingly focus on the complete ***door-to-door*** trip. As far as they are concerned, the individual travel modes being used will be less important than the total time consumed in making the trip.

While trip speed has always been recognized as a competitive factor, the traditional approach of most travel-mode operators is to think only in terms of the point-to-point speed of their particular mode. This has led to the oft-noted anomaly where we may spend a mere hour flying from one airport to another. But we must spend another hour and a half driving from our downtown office to the departure airport, passing through the check-in line and waiting in the boarding lounge—plus another hour and a half collecting our baggage at the destination airport in another city, arranging for our rental car and driving to our downtown hotel. So our door-to-door trip time is four hours, which may be considerably longer than it takes to travel by passenger train from one downtown to another.

Travel-mode operators must become sensitive to the speed of door-to-door trip times for their customers. Among other things, this means learning how to interface more smoothly with other modes.

Second: Most door-to-door trips to move people and goods already involve more than a single mode, and this phenomenon is likely to become more common in the future (unless all passenger travel is to be done in private automobiles and all goods movement is to be handled by trucks, which would be a real disaster). Customers will become increasingly sensitive to the nature of the transfer from one mode to another in making door-to-door trips. They are likely to demand smoother, faster, more transparent and "less fussy" transfers between the various modes they use for a trip. If this demand is not met, many customers may reluctantly choose longer trip times in an effort to avoid transfers. But doing so would mean higher transportation costs for customers and therefore represents a waste of resources that the nation can ill afford.

Unfortunately, intermodal tripmaking has to struggle against entrenched transportation management biases that are narrowly focused on individual modes, plus a chronic lack of funding for intermodal facilities. These are things that must change if the nation's transportation system is to meet the demands of the future.

Third: As customers increasingly demand shorter total door-to-door trip times, the use of two or more modes per trip will become more the rule than the exception. But at the same time, customer impatience with working out the details of modal transfers will also increase. This will provide new business opportunities for firms that offer mobility management services for moving both people and goods.

Mobility managers will provide their customers with integrated travel packages that are custom-tailored to their needs. Using the latest telecommunications technology, they will even be able to revise travel packages on the fly in response to unusual circumstances (like a hurricane striking New York City). And they will relieve the customer of all the messy paperwork involved in keeping track of schedules and reservations for the various transportation modes.

Today's mobility managers tend to be third-party shippers like UPS. But tomorrow is likely to see independent firms whose only business is providing mobility-management services. Smart transportation-mode operators may also add these services as they become aware of the business opportunities they offer.

A SUMMARY OF THE ISSUES

We must keep in mind four issues as we think about how to improve the services delivered by the nation's transportation systems.

1. Transportation facilities exist to support the nation's economy.

Because economic activity generates demand for moving people and goods, it follows that the nation requires adequate transportation capacity in the right places if economic growth is not to be constrained.

Unfortunately, there is much evidence that this need is no longer being met. This is especially true in the twenty largest U.S. metropolitan regions that produce most of the nation's gross domestic product. Worsening roadway congestion, poor interfaces between various travel modes and other transportation bottlenecks in these regions testify to a disturbing lack of adequate capacity on their roadways, public transit systems and rail-freight lines.

Unless we are willing to endure the unpleasant consequences of having to tailor the size of the nation's economy to the limitations imposed by too little transportation capacity, we must renew our focus on making good transportation a national priority.

2. Existing transportation finance mechanisms no longer work.

In both the public and private sectors, the financing mechanisms on which we have traditionally relied provide too little money to meet the needs of the future. What money they do provide is too often allocated among the various modes in ways that reflect the needs of yesterday rather than tomorrow.

In the public sector, for example, the American tradition of tax-based financing of roadways can no longer provide even enough money to prevent existing roadway systems from deteriorating further, much less fund the huge backlog of restoration needs or add new capacity to accommodate growing travel demand.

In the private sector, the U.S. freight-railroad industry is unable to generate a large enough return on its assets to match its cost of capital, despite its generation-long campaign to reduce operating costs by shrinking the size of the nation's rail network. Because the industry is rarely eligible for government capital grants under existing arrangements, this means it is unable to fund the kind of long-needed expansions of rail capacity to accommodate rising goods-movement demand.

One inevitable result of this is that too many goods-movement trips are shifted to roadways, where they add to traffic congestion and accelerate pavement wear.

The obvious solution for funding roadways is to replace the outmoded tax-based financing system with market-based financing by directly charging motorists for roadway use. Fortunately, new technology now makes it a relatively simple matter to charge motorists according to such variables as the number of miles they drive per trip, the size and weight of their vehicles and the time of day when they chose to make their trips.

Under the right conditions, directly charging motorists for trips on limited-access highways in a metropolitan region can generate enough revenue to make the region's entire roadway network fully self-supporting. This would enable us to end the network's dependence on the federal fuel tax and the over-burdened, tax-funded budgets of local and state governments. Instead, we could set it up as an independent enterprise, jointly owned by the public and private sectors, operated according to sensible commercial principles and insulated from swarmy politicians.

Highway user fees can also be channeled as a funding source for other travel modes that directly impact roadway networks. This could, for example, provide financing to support expansions of intercity freight rail networks so that we can move more goods by rail and thereby reduce the burden of truck traffic on roadways.

3. Transportation managers must make SERVICE TO CUSTOMERS their primary focus.

The traditional engineering-oriented focus on "operating transportation facilities" is no longer adequate to meet the more complex and sophisticated demands of the 21st century. It should be replaced by a market-oriented focus on serving travel customers, which is more complicated than it sounds.

The most obvious application of this new focus is in the independent metropolitan roadway enterprises mentioned above. Like the managers of movie theaters, the managers of these enterprises will be confronted with customers paying "admission charges" for access to their facilities. These customers will demand good service in return. In the case of highways, this covers such essentials as always-smooth pavements, modern traffic-control technology and lightning-fast removal of disabled vehicles and weather-related problems that interfere with good traffic flow. It should also involve money-back performance guarantees in the event of significant service failures so that customers can be confident that they are getting their money's worth.

In a larger sense, transportation managers must learn how to embrace the inevitability of change as a source of opportunities to be exploited on behalf of customers. Some of these changes affect the supply side of the transportation equation. They include the range of new technologies coming out of the VII initiative that promise to revolutionize the management of traffic flows, plus the growing recognition of new ways to integrate various travel modes in a more seamless transportation environment. Other changes concern the demand side of the equation. They include increasing insistence by customers on shorter door-to-door trip times, greater value for their travel dollar, more immediate responses to external circumstances that can abruptly disrupt a given trip itinerary, and other realities that the concept of Mobility Management is designed to address

4. **National strategic management for transportation is essential.**

Regardless of the quality of its players, no NFL team can hope to reach the Super Bowl without having an effective strategic plan to guide its performance during the football season. The ability to develop and implement such plans is the most important success secret of such coaching icons as Bill Walsh, Tom Landry and Vince Lombardi.

They understood the importance of beginning each season with a strategic plan that incorporated everything knowable at the time about the performance potential of their own players and how best to exploit it, plus the performance potential of players on the opposing teams and how best to defuse it. And all of this knowledge was written down (along with accompanying tables and diagrams) in thick notebooks.

But they also understood that no pre-season strategic plan is ever carved in stone. It must continually be revised during the season in response to the inevitability of events that can never be anticipated—like injuries to your key players and to those on opposing teams, the unexpected emergence of rookies as star performers and the mystical ability of battered old pros to somehow get it all together one more time as the season develops. No meaningful NFL strategic plan ever has a half-life of more than a week or so. Top coaches know this. They even welcome it because of the fresh opportunities it can bring.

This is the kind of strategic management that can enable us to play heads-up ball as we move ahead with the immense task of restoring the ability of the nation's transportation systems to support (even generate) economic growth. But first we must decide where the responsibility for conducting transportation strategic management should lie.

During World War II, the federal government did strategic management more successfully than any government in history. Led by an exceptionally clear-eyed president and supported by a cadre of talented military and civilian leaders, it developed and implemented strategies to tap the latent ability of the United States to create and manage the well-equipped armed forces that became the key to defeating Germany and Japan—all with astonishing speed. Unfortunately, Washington no longer seems capable of handling this kind of strategic management responsibility. So where do we turn next?

Some experts believe that the answer is to bypass the federal government completely by establishing one or more independent commissions to assume responsibility for national transportation strategic management. Presumably, such commissions would then find ways to snooker the White House and Congress into going along (and paying for) their strategic plans.

Other experts advocate having state governments take over strategic management responsibilities for transportation. Like-minded state governments could band together to plan their transportation futures as "super regions." Since these governments have the ability to generate revenue (and not just from taxes), they would be able to finance their transportation plans on their own. Maybe.

Still other experts believe that the answer could lie in having the nation's major metropolitan regions exploit their growing economic clout to take control of strategic management for transportation—initially on a regional basis, then expanding to incorporate the rural corridors that connect them. The problem here is that metropolitan regions are typically regarded as cash cows for their state governments (not to mention the federal government) and are imprisoned in a form of economic colonialism that keeps most of these governments weak and barely functional. Also, local governments within these regions are so fragmented and contentious that it is difficult to imagine how they could ever cooperate to develop regional strategic plans for transportation. (New York City has the nation's only full-service regional government, embracing five counties that contain eight million people. But it is still only one of several hundred local governments in its vast, three-state metropolitan region.)

It could well be that this question of where strategic responsibility should lie is part of a much larger issue involving the necessary evolution of practical government in the United States during the 21st century. Will the federal government shrink in the direction of something more like what was originally envisioned by the Founders (other than Alexander Hamilton)? Will state governments assume greater functional control over the nation's destiny? Will our major metropolitan regions become quasi-autonomous social entities like China's Hong Kong and manage to gain the kind of political power that matches their economic clout?

These are mega-questions that can occupy the attention of universities and private think tanks for the career lives of their academic thinkers. Meanwhile, we face the much more short-term problem of developing and implementing an effective strategic plan to turn the nation's transportation system into the kind of economic engine we need.

Before it's too late.

INNOVATIVE FINANCE

Strategic Alternatives for Financing the Highway System

The Federal Aid Highway Act of 1956 created the Highway Trust Fund, providing a stable funding source for the nation's highway system, one that was adequate for most national highway needs for about the next 15 years. The Trust Fund receives revenues from a variety of highway-related taxes, approximately 85 percent of which are taxes on motor fuel (gasoline and diesel). The original intent was to generate revenues that bore some relationship to highway use through a budget mechanism that was independent of near-term political forces.

Since the early 1970s, however, the Trust Fund approach to highway finance has encountered a series of structural problems that, over time, have necessitated finding new means of financing the ongoing maintenance and improvement of the nation's highway infrastructure. Starting in the 1970s, for example, the Congress and the Administration began to limit spending from the Trust Fund to help offset the growing federal deficit. This occurred at the same time that a combination of federal regulations and higher fuel prices was stimulating improvements in fuel economy. Over time, the federal highway-finance program changed, and it now has many characteristics of a block-grant program: funds are being distributed as broadly as possible, with little relationship to transportation demand or specific national objectives.

The recently passed Transportation Equity Act for the 21st Century (TEA-21) provides the highway system with a large increase in funding. This increase is the result of two policy decisions: to spend down part of the backlog of cash in the Trust Fund, and to spend all future highway-

---

Adapted from Joseph M. Giglio and Jon Williams, "Strategic Alternatives for Financing the Highway System," TR News *(Washington, DC: Transportation Research Board) 198 (September-October 1998): 34–37, 48.*

related tax revenues on transportation. Since there will be no opportunities in the future for such "house cleaning," any future increases in federal funds for highway finance will require a significant increase in federal taxes.

WHAT'S WRONG WITH THE CURRENT HIGHWAY FINANCE SYSTEM?

The current highway-finance system has three fundamental structural problems:

- Political barriers to raising user taxes.
- Unpredictable revenues.
- Lack of linkage between user fees and highway-system costs and benefits.

Political Barriers to Raising User Taxes

As noted, the Highway Trust Fund was established in 1956 as a user-supported fund. Economists term such fees "benefit taxes." That is, receipts are related to highway use, which in turn is presumed to provide a rough measure of highway benefits. Elected officials and the general public, however, do not view fuel taxes as a direct proxy for the benefits derived from using the highway system. Rather, fuel taxes are regarded as just another tax, and politicians risk losing their jobs if they raise fuel taxes to meet the full costs of the highway system.

The difficulty associated with increasing the fuel-tax rate has resulted in insufficient revenues to maintain, let alone improve, the existing highway system. While TEA-21 funds plus state and local funding should provide an annual capital outlay of about $40 billion in highway finance from 1998 to 2003 (in 1995 dollars), the Federal Highway Administration estimates that $53.5 billion a year is needed to maintain existing highway and bridge conditions. The cost to improve highway, bridge and transit conditions is estimated to average $79.6 billion a year.[1] And even at these high levels of funding, service quality would remain below what it was 20 years ago.

Despite the magnitude of this gap, increases in user fees have been few and far between. During the 1980s, states raised their own fees to match inflation on a regular basis; on average half the states did so each year. By the 1990s, the number of states increasing their gas tax had dropped to an average of fewer than 10 each year. Federal taxes dedicated to transportation had not been raised since the famous nickel increase in 1982 until the enactment of TEA-21, which transferred 4.3 cents per gallon in deficit-reduction taxes into the Trust Fund.

Unpredictable Revenues

In addition to political uncertainty, highway revenues are subject to economic uncertainty involving both demand and supply. The demand uncertainty comes about because fuel-tax receipts are not linked to travel demand alone. Fuel economy varies widely according to the mix and technology of vehicles in the fleet and the presence or absence of regulations.

For example, between 1975 and 1995, fuel economy for passenger cars in the United States increased from 13.5 to 22.6 mpg as a result of the regulation of Corporate Average Fuel Economy (CAFE) standards. Because of these increases in fuel economy, total annual U.S. vehicle gasoline use during the period increased from 99 to 117 million gallons (up 18 percent) while annual vehicle miles traveled (VMT) by automobiles and light trucks increased from 1.2 to 2.2 billion (up 83 percent).

In the long run, advances in technology—especially hybrid vehicles—should make it possible to achieve additional reductions in fuel consumed per VMT, and thus erode fuel taxes still further. Moreover, there is currently no tax linkage to the federal or state highway trust funds for alternative fuels such as electricity and natural gas. Thus, not only does the present system have inherently uncertain demand, but also there is a bias toward lower revenues.

On the other hand, the supply of gasoline in the United States depends in part on international petroleum reserves and production decisions. The supply is also dependent on demand elsewhere in the world. With the current faltering of the Asian economies, there is a global surplus of petroleum, and gas-pump prices are the lowest in years. Lower gas prices, however, increase the demand for travel and gas, and raise revenues.

Lack of Linkage with Highway Costs and Benefits

The current set of highway user fees is not directly related to the costs and benefits of the highway system. Since the primary highway tax is included in the cost of every gallon of motor fuel, it does not have a visible link to use of the system. This lack of linkage encourages inefficient use of and investment in the system. It reinforces the impression that highways are a public good and dissociates the taxpayer from the actual system costs (both direct and indirect).

The result is what economists call the "free rider" problem, with users unwilling to pay the full cost of highway use. The 1997 Federal Highway Cost Allocation Study estimates that federal, state and local user fees paid by all vehicles cover only 80 percent of costs occasioned by highway use.[2] In contrast, consumers pay the full costs of production for most other goods and services.

It is also difficult to track the costs of the highway system since those costs cannot be distributed equitably among users according to how they are incurred. For example, all five-axle tractor semi-trailers weighing between 50,000 and 100,000 pounds pay approximately 6 cents per mile in user fees. Yet these trucks generate costs for highway wear and tear ranging from 3 cents per mile for the lightest truck to 14 cents per mile for the heaviest.[3]

In addition, the current highway financing system makes it difficult to internalize external costs. Examples of external costs include the costs associated with pollution, safety and congestion. Motor-fuel taxes are average taxes and at best are set at levels required to recoup public expenditures. For ease of collection and understanding, the taxes are set at flat rates that at best mirror the average costs of providing the highway infrastructure. Motor-fuel taxes do not encompass the incremental costs of pollution, reduced safety or increased congestion imposed by highway users (and taxpayers) on the system. And while each additional user of the system imposes a greater cost, the increased user charges paid do not cover these higher costs.[4]

Among these uncharged external costs, congestion is perhaps the largest in magnitude and the most important for the nation's economic well-being. Ironically, as adding capacity becomes increasingly difficult because of both funding constraints and limited political will to build new highways, congestion costs rise rapidly as well.

While the current highway-finance system does not track direct and external costs very well, its failure to track benefits is perhaps more disturbing. The value of getting to a business meeting on time or of having a product arrive at the unloading dock when it is needed for a just-in-time shipment is many times the "price" paid through user fees. The lack of anything close to a market test for value received gives departments of transportation and others no incentive to provide better service.

WHAT CAN BE DONE?

We present three options for addressing the above problems with the current highway-finance system. One would work, but is painful and thus unlikely; one represents slight improvements to the existing unsatisfactory policy; and a third offers a long-term solution, but requires leadership and some imagination.

Large Increase in Federal-Aid Program

Until recently, a large increase in federal highway spending might occur in one of two ways:

- Spending the full resources in the Highway Trust Fund.
- Increasing highway user taxes.

TEA-21 appears to have implemented the first option, yet there is still a shortfall in revenues. The agreement to increase spending to equal annual tax receipts, however, also calls for returning part of the cash balance to the Treasury, and future cash balances will no longer accrue interest. Therefore, an old-fashioned increase in federal highway taxes represents the only chance for a large increase in the federal program. Political realities make this option unlikely. Further, such increases would not correct the poor linkage between user fees and costs inherent in the current system.

Muddling Through

Since the Intermodal Surface Transportation Efficiency Act was enacted in 1991, a number of positive events have occurred in highway finance. In dollar terms, TEA-21 is most important, but since 1991 many states have begun to experiment with innovative financing methods, including soft loans and various types of credit support, sometimes using state infrastructure banks.[5] In addition, there has been a modest but mixed history of experimentation with public–private partnerships.[6] Examples include the Southern Connector in South Carolina and the SR 91 Expressway in Orange County, California. Although such approaches are interesting, the pace of evolution is moderate, so that the absolute dollar values generated to date have been limited.

One scenario calls for continued reliance on state and federal gas taxes and other fees, complemented by continued growth in innovative finance. While such a scenario makes for interesting debate over the pace at which evolution might occur, it does not represent a fundamental change. Rather, capital investment is unlikely to make it possible to offset two decades of imbalance between demand and investment. As a result, highway congestion will grow over increasingly longer periods of the day, and there will be deterioration in both highway and economic performance as people and goods experience increasing delays and uncertainties in daily travel.

Paradigm Shifts

Neither higher federal taxes nor muddling through appears to offer a particularly attractive option. Three possible paradigm shifts, none of which are mutually exclusive, offer a better alternative:

- Decentralize to the states and local governments.
- Link payment with use of the system.
- Implement true public–private partnerships.

Decentralize. States and local governments already own most of the nation's highway system and carry out investment and maintenance activities. The federal government serves as tax collector and regulator. Recent years have seen debate over variations on decentralization (with a modest federal role maintained for projects of true national significance or perhaps for cases in which a banker of last resort is needed). Such a shift could encourage a host of new financing approaches by giving states more flexibility in the use of highway trust funds.

For example, the funds could be used as a loan to secure debt financing, as was done successfully for the George Bush Turnpike in Dallas, Texas. With decentralization and greater flexibility, moreover, new management philosophies may evolve, such as increased use of debt financing and leverage, and introduction of public–private partnerships for all aspects of highway construction and maintenance. Indeed, decentralization offers a possible institutional pathway to the next two options.

Link Payment with Use of the System. Electronic toll collection (ETC) has passed from experimentation into standard practice in parts of North America as widespread as Southern California, Oklahoma, Ohio, Florida and Toronto. The E-ZPass program, for example, will encompass a network of toll facilities in Delaware, New Jersey, New York and Pennsylvania; as of September 1998, 2.3 million E-ZPass transponders were in use in metropolitan New York City.[7]

Until now, transponders and related intelligent-transportation-systems techniques have been used primarily to improve traffic flow or to alleviate the difficulties associated with manual toll collection. For the first time, however, transponders are also making it possible to move from an average tax to true use-based fees. Fees can now be set at rates that vary according to the costs imposed by highway users on the system and provide a better measure of the true value received. The result will be fees that begin to act like those in other markets, providing more accurate market signals, as well as significant increases in revenues. Successful experimentation along these lines is already underway. Examples are the SR 91 variable-toll system and the growing interest in high-occupancy toll (HOT) lanes around the country.

Full implementation of such a system would require collection of usage fees for all highway travel—not just tolled facilities. The introduction of transponders, together with the use of readily available global positioning systems, makes this feasible, and would allow variable pricing according to congestion, facility type, vehicle type and time of day. One way of collecting fees through this system would be to use the swipe-card technology that is currently in place at many gas stations. This approach might be called a "variable gas tax"; some drivers would pay a higher and some a lower tax, depending on their use of the highway system. Full implementation of such a system would require both political courage and some time, but this option offers the potential for a far more market-oriented provision of highway services.

Implement True Public–Private Partnerships. Building on the above two changes, the private sector could assume a much greater role in funding, building and operating the system (under public-sector stewardship). Unlike current efforts, such partnering could be implemented on a systemwide or regional basis, rather than piecemeal. Such a partnership approach would recognize that highway services are a public good while enabling private-sector management to offer the most cost-effective provision of services, regardless of whether a road was highly profitable or merely important for social and economic reasons.

As an example consider the Oklahoma turnpike model, whereby some revenue winners support revenue losers, but the system as a whole makes money. The Oklahoma Turnpike Authority built and operates 10 turnpikes to supplement the limited money appropriated for highway construction and maintenance. Turnpike revenues pay all turnpike operating and maintenance costs and pay off the bonds issued to finance turnpike construction. Also, state-maintained roads receive motor-fuel tax money generated by those driving on the turnpikes; the Oklahoma turnpike system generates more than $60 million in state and federal motor-fuel-tax revenues that is contributed to state-maintained roads. Roughly half of the toll revenues collected on Oklahoma turnpikes comes from out-of-state motorists. If tolls were eliminated, the state would have to spend at least $37 to $57 million per year from gasoline taxes to maintain the existing turnpikes, necessitating a tax increase.[8]

A related model is being explored by Argentina as part of an ambitious plan to build a 10,000 km Interstate Highway System. This plan, called Proyecto 10, would combine a national network with private-sector concessionaires to build, finance, toll and operate the system. Up to 300 private concessions would be put out for bid, with competition keeping costs down and service up. The Argentine government would back the plan with energy-tax receipts (10 centavos per liter), payable only after each concession had opened to traffic. In a sense this is a 1990s version of the U.S. Highway Trust Fund. The only missing element is use of technology to link fees with the value of service provided.

CONCLUSION

The U.S. highway system is supported by federal and state gas taxes and other related fees. This set of indirect user fees has paid for the extensive highway network that is the mainstay of the nation's commerce and recreation. One does not lightly change practices that have served well for many years. Yet the Highway Trust Fund approach serves less and less well. It does not directly link fees for use of the system with costs occasioned by users. And it is subject to unpredictable vacillations in revenue that have little relation to highway use. It does not provide enough money to maintain the system.

Despite this erosion, there is no discernible radical imperative to change the current ways of financing the highway system. Yet change in highway finance is going to occur. Change will occur because:

1. The rapid introduction of electronic technologies, such as electronic tolling and automatic vehicle identification, facilitates an efficient, direct means of paying for highway use, and is already being adopted by toll authorities across the nation.

2. States must find new financing mechanisms to support the highway infrastructure essential to their economic growth needs.
3. The introduction of alternative-fuel vehicles requires new and better means of assessing payments for highway use.

These same issues confront governments of all developed and developing nations. The question is not whether there will be change in the way the U.S. highway system is financed. The question is whether we will be positioned to take best advantage of the change when it occurs and to provide a framework that will best suit our local and national interests. Now is the time to develop and implement a strategic plan for reforming the way we finance the nation's highway system.

Notes

1. These estimates are derived from *1997 Status of the Nation's Surface Transportation System: Condition and Performance*, a report to Congress by the U.S. Department of Transportation, pp. 53-55. The estimated cost to maintain conditions is based on the Maintain User Costs estimate, adjusted upward by 16 percent. Dollars cited are adjusted to 1995.

2. *1997 Federal Highway Cost Allocation Study* (Washington, DC: U.S. Department of Transportation, 1997): 21.

3. Ibid, 14.

4. *Transportation Research Board Special Report 246, Paying Our Way: Estimating Marginal Social Costs of Freight Transportation* (Washington, DC: National Research Council, 1996).

5. See Brian Grote and David Seltzer, "Budget Scoring, Highway Projects and Innovative Finance—How the Tail Wags the Dog," *TR News* (Washington, DC: Transportation Research Board) 198 (1998): 15-25.

6. See Joseph M. Giglio and William D. Ankner, "Public-Private Partnerships: Brave New World," *TR News* (Washington, DC: Transportation Research Board) 198 (1998): 28-33.

7. Interview with Frank Pascual (spokesman for the Metropolitan Transportation Authority, Bridges and Tunnels), September 15, 1998.

8. Interview with Neal McCaleb (Director, Oklahoma Turnpike Authority).

INNOVATIVE FINANCE

The Federal Government's Role in a Post-Interstate Era

Part I: History

In 1919, the War Department sent a military convoy of 60 trucks, 258 enlisted men and 37 officers on a cross-country trip from Washington to San Francisco to examine the nation's roads. The "First Transcontinental Motor Convoy of 1919" spurred interest in an extensive national highway network and helped to bring about the Federal Highway Act of 1921. Under the act, Congress and the federal government increased overall funding for roads, but shifted it from local to federal and state programs. An expansion and refinement of the Federal Aid Road Act of 1916, this legislation became the basis for a systematic partnership between the federal government and the states.

But both the 1916 and 1921 legislation discouraged toll financing. The acts prohibited the collection of tolls on roads and bridges financed with federal funds. Congress permitted only a few exceptions to this prohibition. For example, it allowed the use of federal money to construct toll bridges and tunnels and their approaches on the federal highway network. The stipulation was that toll revenues be used to pay for the construction, maintenance and operation of the facilities and that tolls be removed from federal-aid facilities as soon as the original construction bonds were retired.

In 1944, Congress passed another federal-aid bill. It made funds available for cities of over 5,000 people to purchase right-of-ways, and provided for an interstate highway system connecting all the more important highways in the United States. The new network was not to

---

Adapted from Joseph M. Giglio, "The Federal Government's Role in a Post-Interstate Era," a two-part series in The Bond Buyer, *November 6 & 13, 1995.*

exceed 40,000 miles and was referred to as the National System of Interstate Highways.

The demand for more and better roads grew rapidly following World War II. In response, state and local governments began placing greater emphasis on toll facilities and consideration was given to a national system of limited-access toll highways. But interest in tolls as a primary source of road funding soon began to wane.

The Money Tree

In 1956, President Eisenhower—who, as a young Army lieutenant, had helped write the report for the "First Transcontinental Motor Convoy of 1919"—signed a new piece of legislation that resolved the key financing issue facing the U.S. highway system.

The Federal Highway Act of 1956 authorized $25 billion for a 10-year effort to construct 40,000 miles of interstate highways. An early example of innovative financing, the program was to be funded not through general revenues, but through a Highway Trust Fund, whose revenues would come from new taxes on the purchase of fuel, automobiles, trucks and tires. Also, federal money for the interstate system was made available on a new sharing arrangement under which the federal government would cover 90 percent of costs, while states would foot the remaining 10 percent.

The Act also provided for some toll roads built prior to 1956 to be included in the interstate system as long as the toll facilities were removed at the earliest opportunity. In addition, it granted exceptions to certain toll roads on a case-by-case basis. Although many of these exceptions required that all federal funds be repaid, the repayment did not include an interest charge, and the funds were credited back to the state's unused balance of federal-aid funds.

Many more such exceptions were granted under the Surface Transportation Act of 1978, creating strong economic incentives for states to retain toll facilities. This new Act represented a departure from the early federal policy of strict opposition to toll roads.

Primarily as a result of the creation of the Interstate Highway System, public spending for highways grew dramatically after 1960.

But at the same time, two big problems emerged. First, even though the nation's highways were in serious disrepair, federal aid was oriented primarily toward new construction. Second, highway-construction costs were rising rapidly due to inflation, while federal revenues dedicated to highways—primarily the four-cent-per-gallon, fixed-rate gasoline tax—were declining as a result of lower gas consumption by drivers purchasing more fuel-efficient cars to cope with higher fuel prices.

Congress responded to the first issue by amending the Federal Highway Aid program in 1974 to permit federal funds to be used for resurfacing, restoration, and rehabilitation and setting aside trust fund money for this purpose in 1976. The second problem of declining federal revenues was not addressed until Congress passed the Surface Transportation Act of 1982. As part of the Act, Congress raised the two primary taxes that provide revenues to the Federal Highway Trust Fund. The excise tax on gasoline was raised from four cents to nine cents per gallon; all but one cent was dedicated to highway use. At the same time, highway-use taxes levied on trucks were raised significantly.

Meanwhile, the federal government continued to move away from its opposition to toll roads. In 1987, with the passage of the Surface Transportation and Uniform Relocation Act, Congress provided for federal participation in seven "pilot" toll projects, subsidizing 35 percent of construction or reconstruction costs in an effort to more effectively finance these needed facilities.

In sum, by 1987, the federal policy opposing toll roads had been gradually eroded, and toll bridges and tunnels had been constructed with federal revenue.

The Cleavage Within

From a broader historical perspective, we have come full cycle in financing highways. At the beginning of the 20th century, almost all of the country's roads were supported out of the general funds of local governments and by private enterprise. Federal and state governments did not participate vigorously in road building until well into the first quarter of the 20th century.

As the financial responsibility for highway activity changed, so did the philosophy of paying for highway services. During the first part of this century, highway services were paid for by the general taxpayer. The rise of gasoline taxes and motor-vehicle registration fees reflects the philosophy that the beneficiary pays. At all levels of government, highway users help pay for highways through the gasoline tax.

Since the gas tax was adopted in 1956, the Highway Trust Fund has financed the greatest road-building program the world has known. This achievement is a tribute to the political power of the partnership arrangement between the federal government, the states and the consortium of automobile, oil and rubber industry executives and senior highway bureaucrats that form the road gang.

Over the last 15 years, this partnership has been eroded by the imposition of obligation ceilings, delays in federal appropriations from the Highway Trust Fund, attempts to divert fuel revenues to other purposes and the increase in federal mandates such as the 55-mph national speed limit, mandatory drinking-age laws and national truck size and weight standards typically accompanied by the threat to withhold funds or project approvals.

Taking Stock

As we approach the end of the "American Century," only a few problems in our society are more challenging and intriguing than highway financing.

Economic changes, including growth in the global economy and shifts in corporate structure and ways of managing operations, have created a new set of demands on highway infrastructure. To compete globally, the private sector must have better roads. Still, we have more than $300 billion in backlogged bridge and highway needs; about 25 percent of our major highways—234,000 miles—are in "poor" or "fair" condition and we need to find ways to attract additional capital as one of the means to bridge the gap between highway needs and dedicated funding. The bottom line is we need to invest an additional $10 to $20 billion annually just to run in place.

These new demands have arisen amid promising changes in the landscape of transportation policy and finance. An increasing awareness of environmental problems, including land use and air-quality issues, has spurred government action to curb vehicle congestion and limit the effects of air pollution. Moreover, the legal and regulatory framework for transportation decision-making has changed markedly in the last several years. The Intermodal Surface Transportation Efficiency Act (ISTEA) and the Clean Air Act Amendments of 1990 have combined to create a new framework for transportation planning and finance.

New technologies, including those that fall under the Intelligent Transportation Systems rubric, have also created new capital-financing opportunities for addressing highway needs.

But all this change is taking place even as government budgets get tighter and traditional highway-financing mechanisms are under siege. Diversion of highway revenues continues to be an issue.

Again in 1993, an additional 4.3-cent gasoline tax was passed to be used for deficit reduction. Given the current mood in Congress and the popular resistance to new taxes, highways will continue to compete for appropriations with welfare and healthcare and continue to run the risk of becoming just another government program.

PART II: THE NEW FEDERALISM

The 1980s witnessed a dramatic realignment in the federalist system not seen since the 1950s. The difference this time is that the initiative came from below, with Congress acting at the insistence of the governors. Under the new federalism concept introduced by President Reagan, a sharp slowdown occurred in the rate of growth of federal grants to state and local governments. After a brief interruption, the notion of decentralization has taken hold again.

But while Congress is willing to give states more responsibility, it is also proposing that they manage it with a whole lot less money. Dubbed "devolution," this new approach holds that states, given their closer proximity to the people they serve, can allow more local participation and less expensive services. The public's resistance to tax increases (including the gas tax) may well reflect growing support for tolls and privatization initiatives. The trouble is that if states are fiscally unprepared to pursue the twin goals of economy and efficiency, service quality and reliability could suffer.

The Charms of ISTEA

The Intermodal Surface Transportation Efficiency Act of 1991 represents a dramatic shift in transportation policy. The 1991 reauthorization provided an opportunity to reshape the surface-transportation agenda and signified the conclusion of an era that began in 1956 when the Interstate Highway program was originally authorized.

While the ISTEA six-year authorization of more than $155 billion represented a new high for transportation, the centerpiece of the legislation is the expanded level of flexibility offered to state governments. ISTEA provided new revenue options in the form of tolls; a waiver of state match; an ability to use toll revenues that are generated and used by public, quasi-public and private agencies as the match to federal highway funds; and the ability to recycle transportation-related grant funds through a state highway revolving fund, to be used for additional highway projects. ISTEA also allowed greater private involvement in building, maintaining and operating toll roads and bridges and allowed federal funds to be spent on private toll roads.

Initially, these new financing opportunities did not circulate as common currency because they represented a departure from traditional financing methodologies. States expressed concerns about the their complexity and about the constraints imposed by state and federal laws and regulations.

A Different Model

To address these issues, the Federal Highway Administration announced in March of 1994 the "Innovative Financing Program" (TE-045), which made it easier for states to take advantage of the ISTEA financing opportunities as well as some of the other financial tools developed by the private sector and other states.

FHWA's objective for TE-045 was to identify existing transportation-financing barriers and to encourage innovative financing mechanisms that would increase highway investment. This initiative asked states to identify specific projects that would be advanced through new ways of financing and that hold the potential to increase transportation investment.

Fortunately, the TE-045 initiative has drawn major interest and does not represent the kind of pallid government program usually gathering dust among thousands of others. Acting as the catalyst in leveraging capital to bridge the transportation-financing deficit, FHWA received proposals from over 30 states with projects that cost a few hundred thousand dollars up to $1 billion. FHWA has accepted about 45 of these proposals. The projects are diverse and financially creative. Some are intermodal in nature, reflect geographical diversity and include projects in rural, suburban and urban areas.

In sum, the TE-045 Innovative Financing Program represents an important first step to move the current transportation-financing process from a single strategy of federal funding on a grant-reimbursement basis to a diversified approach providing essential new resources to state DOTs.

Federal Initiatives

In addition to helping accelerate new projects, the state submissions covered financing ideas that help inform legislative alternatives, especially in an environment where the administration and the new Congress have made increasing state financing capacity and flexibility a primary objective. Indeed, based on the lessons learned from TE-045, state infrastructure banks were included in the restructured DOT program and the president's fiscal 1996 budget. Still further, the Senate version of the National Highway System bill reflected several of the innovative financing ideas developed through TE-045, including the following:

- Increasing flexibility in allowing bond interest and issuance costs to be eligible for federal reimbursement.
- Broadening the current ISTEA toll provisions, which allow states to loan amounts equal to the federal share of a project to a public entity (toll authority) as long as a dedicated revenue source (tolls) exists to support the project. The new amendment would extend this to include projects for non-toll facilities backed by dedicated revenues such as excise taxes, sales taxes, motor-vehicle-use fees, real-property taxes, etc.
- Allowing materials, right-of-ways, engineering plans, and donated funds to count toward the state's matching share.

Here again, the new financing techniques are an effort to leverage scarce transportation resources and attract capital from nontraditional sources.

Entre Nous

While so far no one has raised the issue of abolishing the federal Department of Transportation, the fundamental question remains: What should the federal government's role be in the post-Interstate era? The current fiscal debate suggests two basic directions. One is to remove trust funds from the budget. Another is to turn back all or part of the revenues to state governments.

The Highway Trust Fund is a convenient starting point for considering various strategic options under the new federalism. Since 1956, it has been the driving force behind federal-highway financing. But the Highway Trust Fund has been part of the unified federal budget since 1969. In retrospect, this budget change was like handing over the keys to the wine cellar to the town drunk. This has made the trust-fund balances, which currently aggregate about $18 billion in illiquid Treasury IOUs, available to subsidize the general-fund budget and undermine the trust-fund concept.

It is too easy to overlook the fact that the Highway Trust Fund resources are fundamentally user-fee revenues held in trust for state use. It is for this very reason that the chairman of the House Transportation and Infrastructure Committee, Bud Shuster, R-Pa., introduced legislation this year to remove the federal Highway, Aviation, Waterways and Harbor Maintenance trust funds from the unified federal budget. The objective of this legislation is to insulate transportation trust-fund revenues from diversion, the bête noir of the highway industry.

The administration's proposed fiscal 1996 budget, for example, would open up the Highway Trust Fund to pay for Amtrak, mass-transit operating subsidies and all U.S. Department of Transportation administrative and operating expenses.

Looking back, as a malign afterthought, the issue of restoring trust in the Highway Trust Fund and fending off constant raiding parties on the fund may have been a minor issue if Eisenhower had accepted General Clay's recommendation to issue bonds to pay for the Interstate Highway System, which would then be repaid through the gas tax. Both lenders and the highway industry could place greater reliance on the bond indenture and market discipline to protect the Highway Trust Fund from political interference and special mandates.

Another perspective suggests giving responsibility back to state governments and concurrently ending federal excise taxes. This is appropriately called the turnback approach. This scenario would offer state and local governments the opportunity to increase their own excise taxes without any apparent change in cost to the consumer. Of course, this assumes that states would retain the full federal gas-tax turnback and earmark all the revenues for highway purposes. Lots of luck. Diversion may even be a more serious problem at the state level in the service of simple politics. The benefit of this shift would be to posit revenue and spending authority in the hands of government closer to the project. This approach is also consistent with state government's lead in economic development. Additionally, the turnback approach would diffuse the issue of state-to-state fairness.

But there is room for middle ground here. Most observers agree some continuing federal program will be necessary to serve programs of clear national need, such as the Interstate Highway System. Here the federal government extensively funded a program because of its direct responsibility and interest. And the federal role in highway expansion goes beyond the Interstate Highway System.

ISTEA, in 1991, established a 155,000-mile National Highway System that also qualifies for federal assistance. A coalition of highway interests developed this new goal for the federal highway program as the successor program to the Interstate Highway System.

Recently, the U.S. Senate passed S 440, legislation that designated the new National Highway System as the focal point for future federal highway investment. This system includes the entire Interstate Highway System, the national-defense highway network and other highways of national and regional significance that have been identified by the federal Department of Transportation in cooperation with the states. It would ensure that $6.5 billion in already authorized funding could be made available to states in fiscal 1996 and 1997 to support Interstate maintenance and National Highway System route improvements.

Still, even with the passage of the new national system, it is feasible to implement a modified turnback approach by adjusting the revenue-generating capacity of the Highway Trust Fund or even considering a block-grant approach to encompass the remainder of the federal-aid system. Here again, this represents a graceful way to start to make an exit. Taking this direction, with the accompanying revenue pass-through, is a good starting point for recognizing that the states should occupy the central role in highway responsibilities and for aligning the funding mechanism to enable them to meet their needs.

These issues are far from settled, and we may even witness a real debate about privatizing the Highway Trust Fund. Now that the FHWA has brought innovative transportation financing out of the closet, the debate over the use of these new financing tools and the future federal role should be spirited and joined as part of the ISTEA reauthorization process. But one thing is for certain: no single strategy or financing tool is going to build and sustain a world-class highway system. Even if the larger issues about the proper role and size of future federal involvement are not addressed, states can benefit by using these new financing tools.

The New Paradigm in Financing Transportation Infrastructure

With the recent proliferation of innovative financing tools and the accompanying rosy rhetoric heralding their arrival, there is a temptation to deliver a eulogy in honor of traditional transportation finance. But some might say doing so is more akin to staging one of those buried-alive horror scenes from Edgar Allan Poe.

Still others might suggest that instead of trying to bury traditional financing alive, proponents of innovative financing have come to orate over an empty coffin. Maybe traditional financing is not dead but only temporarily missing, which presents something of a *corpus delicti* problem. Still, the question remains: is traditional transportation finance, like virginity past the age of 20, a dead letter? Fortunately, the answer to this question is "No." Innovative financing is not about to dethrone traditional financing.

What is the new paradigm in financing transportation infrastructure and why is it happening? The new paradigm is evolving in response to the concern that traditional transportation funding will not provide sufficient resources to correct the underinvestment in transportation infrastructure.

Widening Gap

Federal, state and local governments recognize the importance of transportation investments to the health of the economy. Yet the gap between the demand for transportation-infrastructure investment and governments' ability to finance these investments is growing. The U.S. Department of Transportation (USDOT), for example, estimates that annual spending on highways, bridges and tunnels would have to

---

Adapted from Joseph M. Giglio, "The New Paradigm in Financing Transportation Infrastructure," ITS Quarterly *4, no. 3 (Summer 1996): 50-55.*

increase by about $20.8 billion just to sustain existing facilities, and spending would have to increase $36.2 billion to realize any net improvement.

At the state and local level, infrastructure-investment needs compete with other demands for increasingly scarce public money. At the federal level, the Highway Trust Fund is used to paper over the federal budget deficit. Viewing the treatment of the Highway Trust Fund over the years is like watching the air slowly leak out of a tire.

Given the current mood in Congress and the popular resistance to tax increases, transportation-infrastructure funding will continue to compete for resources with social-infrastructure programs. This is the predicament of contemporary transportation-infrastructure financing. So innovative finance is a rapidly developing field that has emerged in response to the environmental reality: traditional transportation financing is under siege.

The fundamental reality of the new paradigm is that while the traditional reliance on motor-fuel taxes and federal-aid highway funds will continue, the resources required to fill the funding shortfalls will draw on new financial tools such as privatization, flexible management of federal funds, new revenue sources, credit-enhancement mechanisms, leveraging tools and state infrastructure banks.

It is reasonable to conclude that as the gap between transportation-infrastructure resources and transportation-infrastructure needs widens, it will open the door to a larger role for capital markets in financing transportation investments.

Creative Energy

Throughout the country, transportation policy makers are changing the way they think and work with finance issues. As we approach the millennium, the necessity for new revenue sources and capital intensifies. Transportation policy makers are engaged in a transformation that is leading to the new paradigm in financing transportation infrastructure.

Traditional transportation financing, once a static tapestry of carefully cultivated relationships between the states, the federal government and industry, is displaying a new burst of creative energy.

Thanks to the efforts of a relatively small number of people who made a major contribution, innovative financing tools have taken on the currency of a full-blown cliché. For some transportation officials, using them has become a habit, maybe even a tic.

The Intermodal Surface Transportation Efficiency Act (ISTEA) of 1991 provided states with new leveraging tools and greater flexibility to meet their transportation investment needs. These opportunities were expanded under the Federal Highway Administration's Innovative Finance Test and Evaluation (TE-045) program.

Standing out prominently amid this sea change in policy initiatives is, of course, the recent passage of the National Highway System Designation Act of 1995 (NHS). This legislation incorporated many of the innovations developed by states under the TE-045 innovative financing program and created numerous opportunities for state transportation agencies and authorities to leverage federal and state transportation resources.

Tools for ITS

The flexibility permitted by ISTEA and subsequent statutory and policy changes at the federal and state levels encourages a broader, more comprehensive and intermodal approach to transportation. This expanded approach includes the use of these new financing tools to deploy Intelligent Transportation Systems (ITS) technology. For the uninitiated, ITS technologies hold the promise of improving the efficiency and safety of the federal-aid highway system and provide a cost-effective alternative to expanding and managing physical capacity.

Finally, the climate of innovative finance encourages states to work with new, non-traditional kinds of project sponsors, including independent authorities, private developers and local government entities. The central idea is that federal transportation policies are quickly evolving away from a single strategy of grant reimbursement.

Programs Underway

Efforts to attract private capital and to involve private-sector sponsors are already underway. California's private SR 91 express lanes (using variable tolls) opened last December. Arizona, Delaware, Minnesota, South Carolina and Virginia have active solicitation for private-sector investments in transportation infrastructure. Colorado, Florida and Oregon have passed enabling legislation encouraging private-sector investment. The Texas Department of Transportation made a $135 million ISTEA loan to leverage construction of a $900 million toll road. Ohio proposes to pay for one-third of its $200 million annual expansion in capacity with the new innovative financing tools.

Better yet, the NHS legislation directed USDOT to designate 10 states to implement pilot State Infrastructure Bank (SIB) programs. At the core of the SIB concept is a revolving loan fund that can provide states with a perpetual source of capital for transportation projects. SIBs hold the promise of combining the various innovative financing tools under one entity to focus a state's transportation investment strategy while at the same time providing for additional leveraging of scarce transportation resources.

USDOT has selected Arizona, California, Florida, Missouri, Ohio, Oklahoma, Oregon, South Carolina, Texas and Virginia for the pilot programs. Projects identified in the pilot SIB program range from Ohio proposing a $30 million construction loan to the Butler County Transportation Improvement District to a proposal for the Oklahoma Turnpike Authority to finance a $200 million project.

These are just samples of the emerging direction in financing transportation-infrastructure investment.

New Tools

These new core concepts and innovative financing tools will be essential in developing the financial toolbox needed to bridge funding shortfalls. The overarching objective of the new paradigm is to offer transportation policy makers a new way of doing business based on flexibility, leveraging and a greater orientation toward capital markets. These new tools are applicable to a wide range of transportation-infrastructure financing issues.

These new tools are also highly interrelated and can be used in combination to finance a specific project. But they must be integrated with the strategic planning and operating activities of state transportation agencies, independent toll-road authorities and metropolitan planning organizations if they are to contribute to expanding capacity and correcting underinvestment.

In this way, innovative financing is not an end in itself but a contributor to the improved performance of transportation systems as a whole. When used in a complimentary fashion, the various innovative financing tools offer the potential opportunity to build more projects with fewer dollars and get projects in the ground sooner.

They may also help facilitate the funding of large-scale capital projects through the use of project finance. Through this mechanism, a project is able to attract funding entirely based on the value of the new asset to be built rather than on the strength of the entity that will own and operate the asset.

Project financing has been used to finance energy exploration, oil tankers, refineries and electric-generating plants. Among the distinctive characteristics of project finance is that typically creditors do not have full recourse against the project sponsors. Lenders provide financing for a project against the strength of long-term contracts from credit-worthy buyers that will purchase the output at prices sufficient to cover debt service.

The increased flexibility and leveraging opportunities embedded in the new paradigm may promote the use of project finance by exploiting the generous loan provisions under section 1012 of ISTEA as expanded by section 313 of the NHS act. These involve relaxed restrictions on income-generating activities, the availability of credit enhancements and the use of flexible federal-aid reimbursement rules.

State transportation agencies can now begin to develop financing arrangements with private developers to access the capital markets which have historically provided project financing, such as the insurance companies and commercial banks, to facilitate the financing of large-scale, high-priority transportation-infrastructure projects. A SIB is a powerful and convenient mechanism to bring to bear on large projects all the innovative financing tools available under the new paradigm. The innovative financing tools are available to non-SIB states as well.

Why Create a SIB?

The SIB concept was developed to assist states in their efforts to leverage current funding and produce additional funds, both public and private, for investment in highways, transit, rail, intermodal projects and new technologies, including those under the Intelligent Transportation Systems rubric.

So why should a state consider creating a bank? Let me count the ways. First, it allows states to leverage existing resources. They can build more projects with fewer dollars and accelerate project construction, especially for projects where economic benefits can be identified and captured. This approach ameliorates the impact of inflation on construction costs and helps to realize project benefits earlier. These benefits will in turn generate jobs, private-sector income and tax receipts.

Second, by offering an array of financing tools such as low-interest loans, refinancing, subordinated debt instruments and construction financing, the SIB can tailor financing packages to meet specific project needs and provide increased flexibility. Closely related, infrastructure banks can facilitate projects that are financially tenuous by providing credit support through lines of credit or bond insurance. Equally important, the availability of a menu of financing tools coupled with the ability to offer subordinate debt financing can attract both non-traditional private capital and local government resources, further enhancing a state's ability to husband scarce transportation resources.

A third consideration in creating the bank is the opportunity for states to develop their own self-renewable, insulated source of future capital. Simply put, the state banks have the ability to recycle resources by re-loaning funds as they are repaid. Effectively, the repaid funds become state resources. This means that, in addition to increased leverage and additional flexibility, states have the opportunity to develop and control their own source of capital.

Finally, for states that can work past their deep and abiding distrust of debt and bankers, an infrastructure bank can gain greater leverage by issuing debt against the bank's capital so that even more funds can be made available for lending. This accelerates the recycling of loan repayments, increases the magnitude of available transportation resources and provides for a larger financial canvas with which to work. Dissenting opinion is never in short supply in the transportation business, but the only thing I can find wrong with infrastructure banks is that it is a sin *not* to use one.

Not Limited to Highways

Actually, that was the penultimate point.

The final point is that infrastructure banks are not limited to highway projects. They are designed to provide assistance to all eligible Title 23 projects, including transit, intermodal facilities and certain rail projects. This provides states with the opportunity to take an integrated approach to tying transportation planning and finance together to create an integrated intermodal transportation network.

This new, integrated approach reflects the contribution of ITS technologies to provide additional transportation capacity. If a primary objective of innovative finance is to maximize the ability of states to leverage transportation resources, then cost-effective ITS technologies must be considered an essential component of any strategic plan to provide and manage additional capacity, particularly when it is understood that we function in a mercantile world, with limited resources. The ITS community may want to consider taking advantage of the TE-045 program to propose the application of the innovative financing tools to leverage resources for specific ITS deployments.

National Needs, Local Situations

In the new paradigm for financing transportation infrastructure, a key challenge is to tailor the various core concepts and innovative financing tools to reflect the discrete differences among the states. Each state has a unique political-legal, socio-cultural, economic and financial environment, which presents possible opportunities and threats that significantly affect the viability of the new paradigm. This situational specificity is one reason the new paradigm cannot really be defined in a final way. No one way of doing business is best for each state.

The transition to new financing patterns may be a slow and arduous process. For many states, innovative finance is terra incognita. For example, in certain states legal limitations will serve as a serious barrier to change. Some states place restrictions on state bonding authority that will have to be modified before any form of transportation related debt issuance can occur. Other states do not permit the lending of state credit to private entities, and in others, concerns about increased debt issuance actually border on hysteria. In most, legislation will be needed to authorize the creation of SIBs.

The challenge is how to expand and sustain the opportunities of the new paradigm and to reconcile the potential of ITS technologies with innovative financing tools. If not, much of what passes for innovative financing will be little more than a collection of anecdotes, rules of thumb and crude manipulations of financial leverage—just waiting for the final nail in the coffin.

INNOVATIVE FINANCE

Financing Transit in Large U.S. Cities

The U.S. transit industry faces a series of profound economic, social and political challenges that force it to consider alternative ways of providing urban transportation services. Without significant structural change, the industry's ability to survive in its current form has to be questioned. This paper focuses on financial and related operational changes that can provide practical help in the near term. Many of these options have been selected, however, for their potential to assist in resolving long-term problems as well.

How serious is the problem? Despite more than $100 billion in federal investments over the past quarter century, transit continues to lose market share to the private automobile. For example, between 1980 and 1990, the share of those driving alone to work increased from about 64 percent to about 73 percent, while the fraction of commuters who traveled by carpool declined from about 20 percent to about 13 percent. Commuters using transit decreased from 6.4 percent to 5.3 percent during the same period. Transit is now the mode of last resort for most urban travelers, and it lost business in both absolute and relative terms between 1980 and 1990. Any private business faced with these losses would be disinvesting.

Five long-term trends appear to account for these problems:

- Increases in personal income mean people can afford to select higher-quality, more expensive forms of transport. This movement to

---

Adapted from Joseph M. Giglio, "Financing," in Transportation Issues in Large U.S. Cities: Proceedings of a Conference held in Detroit, Michigan on June 28-30, 1998, *111–125 (Washington, DC: Transportation Research Board, National Research Council, 1999).*

quality can be seen across all markets and reduces the size of transit's captive market.

- Where we work and live is more dispersed. Most jobs and homes are in the suburbs, with a lower density of trips between specific locations—ironically, with greater general congestion. Traditional transit systems have been built along fixed routes and fare poorly when demand is dispersed.
- Changes in lifestyles mean that the classic home-to-work trip no longer dominates urban travel. Two-income families and single-parent families are more common than the traditional nuclear family with its single breadwinner. As a result, fewer trips have a single purpose, and the amount of "trip chaining" (the process of linking multiple destinations into a single trip) has grown. Together, these trends favor modes with greater flexibility, such as the automobile.
- The pace of business has also changed. Reliability and predictability are now more important than cost. The move toward "just-in-time" systems affects workers as much as business.
- Transit remains the last major mode whose market is protected from commercial competition (although in effect personal vehicles compete directly with transit). This lack of competition for provision of transit services has reduced openness to new ideas.

In addition, local and regional transit agencies face many mundane but still daunting challenges. A non-exhaustive list of challenges faced by many transit agencies includes the following:

- Many transit agencies need to find funds to replace federal operating assistance that is no longer available.
- The costs of travel to which consumers are most sensitive—such as the costs of fuel and parking—are either at all-time lows or subsidized by employers and developers. The vast majority of new jobs in the suburbs include free parking.
- Transit agencies are increasingly required to support the mobility needs of society's most disadvantaged populations.
- Experience has shown that transit can attract single-occupancy-vehicle commuters, but only if it enjoys travel-time and travel-cost advantages. Without preferential treatment for buses, including the

use of new technologies, transit can offer only limited incentives to "choice" riders.

Together, these challenges create an environment in which it is increasingly costly per passenger-mile to operate, increasingly difficult to compete with private vehicles in the service dimension, and increasingly challenging to find funds to maintain service levels.

At the same time, transit agencies have significant opportunities, both technical and strategic, that may help them meet these challenges, including:

- More flexible federal funding for transit projects under both the Surface Transportation Program and the Congestion Mitigation and Air Quality Program, and proposals to allow greater flexibility between capital and operating funds.
- Transit agency representation on metropolitan planning organization (MPO) boards, giving transit agencies an institutional role in regional transportation planning.
- Increased congestion on the nation's roads, making transit an increasingly attractive alternative to the stress of single-occupancy-vehicle travel.
- Growing public pressure to reduce pollutants, including greenhouse-gas emissions, making transit systems an increasingly attractive component of air-quality and global-climate-change strategies.
- Growing power and adaptability of technology via intelligent transportation systems (ITS) to improve transit efficiency and raise service quality.
- Growing commitment to making suburban employment accessible to urban residents, thus increasing the role and visibility of transit systems as a mechanism for meeting social goals (while offering the potential to generate revenues on reverse commutes as well as on primary commute routes).
- Development of new financing tools such as State Infrastructure Banks (SIBs).
- Growing recognition of paratransit as a way to complement and supplement traditional transit operations.

Together, this non-exhaustive list of opportunities focuses attention on how transit systems can help society meet many of its important goals, be increasingly competitive with other travel modes, take advantage of new financing tools, leverage private-sector efficiencies and do what it does better.

This article traces historic and expected funding trends and describes financing tools and approaches that can help transit systems confront the challenges they face.

When considering the usefulness of innovative financing tools, four ideas are important to keep in mind.

1. To implement most innovative finance tools and to leverage private-sector participation, it is crucial to be able to think like private-sector partners. These efforts will be most successful, in turn, when they are combined with a good understanding of the direct and indirect economic benefits provided by transit service.
2. In most cases, innovative financing tools and private participation will help reduce reliance of transit system operators on public funds, not replace public funds.
3. Transit system operators must sit at all of the funding tables, not just those to which they are accustomed. As transportation funding sources become more flexible, as the transit environment becomes increasingly multimodal, and as states and MPOs continue to jockey for influence, transit operators must not let innovative finance opportunities pass them by because of a misplaced focus on traditionally reliable revenue sources. New arenas for transit participation could include chamber of commerce committees, business roundtables and economic development agencies.
4. It is important that transit operators not become caught up in the technical details of innovative finance mechanisms. Such mechanisms are relatively straightforward to write up, and many public documents already describe them in detail. Inevitably, though, these documents present hypothetical scenarios for the implementation of such mechanisms. These scenarios are often coupled with detailed discussions of risks inherent to each mechanism—political risk (e.g., benefit-assessment districts), market-growth risk (e.g., joint development potential), currency-exchange-rate risk (e.g., cross-border leases), and so forth. In this environment, it may be easier for transit-agency officials to fall back on the tried and true methods of raising funds. But it is important to emphasize that the arena of uncertainty is where hypothetical situations can be devel-

oped into reality and where innovation takes place. As a result, the most innovative aspect of any finance tools used by transit agencies will not be the tools themselves, but the creativity, iconoclasm and persistence that transit agencies show in implementing them.

Historic and Expected Trends for Federal, State and Local Operating Funding

In 1981, annual federal outlays for transit reached about $5.4 billion (all figures are in constant 1997 dollars unless otherwise noted). After that date total federal outlays began to decline. Total outlays declined to about $3.2 billion in 1992, and operating assistance fell from about $2.1 billion in 1981 to $800 million in 1995 before being eliminated for all but the smallest systems in 1997.

In response to these pressures, transit expenditures have increased sharply at the state and local levels, rising from about $9.7 billion annually in 1982 to slightly more than $17 billion annually by 1994. This increase has several principal components, including replacement of federal funds, operating support for transit new starts for which the federal government has provided only capital assistance, and expanded service in suburban jurisdictions.

As a result of these shifts in the sources of transit funding, fare-box revenue now represents about 40 percent of total transit system receipts, state and local funding sources provide a little over 20 percent each of system receipts, and non-fare revenues (including such receipts as those from advertising, interest and joint development) generate about 13 percent of receipts. Only about 4 percent of all transit system operating income comes from federal sources.

No major change in the magnitude of federal funding for transit operations is expected. To the extent that there are relative changes in levels of state and local support for transit operations, such changes are likely to be determined by the responses of different transit agencies and states. Transit agencies are divided generally into those receiving the majority of their nonfederal public funding from states (e.g., MTA in New York City) and those receiving the majority of their nonfederal public funding from local sources (e.g., MARTA in Atlanta).

As a result, transit agencies can be grouped into two categories for purposes of considering opportunities for innovative funding and financing: those whose greatest opportunities are in the state and "other revenue" categories, and those whose greatest opportunities are in the local and "other revenue" categories.

State Sources of Transit Operating Assistance

Of the systems that rely primarily on states as their major source of non-federal public funding, the largest fraction, slightly more than one-third, is provided by the "other" category. This category predominantly includes general funds but could include state lotteries, toll revenue set-asides or state special taxes. Gasoline and sales taxes provide slightly less than one-fourth each of all state funding and are followed by income taxes, which provide almost all of the remainder. Property taxes represent a mere 1 percent of all state-provided transit operating assistance.

Local Sources of Transit Operating Assistance

For systems that rely primarily on local governments as their major source of non-federal public funding, most support is provided by dedicated sales taxes. This category provides about 80 percent of all such funds. The remaining 20 percent of local funds is composed of property taxes (8 percent), "other" revenues (7 percent), income taxes (3 percent), and gasoline taxes (2 percent). Many local governments provide for transit out of their general tax proceeds and do not have dedicated transit taxes.

Sources of Capital Investment

In 1995, federal sources provided 51 percent of capital investment in transit systems, state sources provided 13 percent of investment, and local sources provided 36 percent. States that are unusual in terms of the support they provide to transit capital investment include California, New Jersey, the District of Columbia and selected others.

GETTING TRANSIT AGENCIES TO THE TABLE

As transit operators consider how to take advantage of the funding sources available to them, and as they consider how to use these funding sources creatively, it is important for them to take the first step: getting to the table when funding tools are being developed, when funds are being allocated, and when new facilities are being prioritized.

The importance of this principle is seen in the degree to which state highway funds seem disproportionately targeted toward rural and lightly populated areas. For example, 60 percent of the New Hampshire residents live in cities with populations of more than 50,000, but these cities receive less than 7 percent of state funds used to build roads. The same goes for funds used to maintain roads. In New York, these cities are home to 97 percent of the state's population but receive only 11 percent of state highway maintenance dollars. For whatever reasons they occur, such disproportionate funding levels indicate that representatives of urban constituencies and providers of service to these constituencies could benefit from spending more time advocating for their transportation interests, whether highway or transit related.

Two other examples illustrate the importance of "being at the table."

1. In the 1970s, the Georgia Department of Transportation (GDOT) reconstructed the radial interstate system serving Atlanta. One major component of the effort was to build left-side high-occupancy vehicle (HOV) ramps for future use when HOV was implemented. For the 1996 Olympics, GDOT restriped the existing lanes and narrowed the shoulder areas to create a new inside lane on 1-85 and 1-75 for HOV 2. While the HOV ramps were being planned and constructed, MARTA was in the middle of building its heavy rail subway system. Three of the four major legs of the rail system parallel GDOT's HOV system now in place.

While both the rail system and the new HOV system are worthwhile endeavors, they are not coordinated and they operate independently. MARTA does not run express bus service on the interstate system, preferring to run local feeder bus service to the more efficient rail system. As a result, MARTA is not running any buses on the GDOT HOV system, and there is no sharing of park-and-ride facilities. Had MARTA officials participated in the roadway decision-making at the right level and the right time, they may have made a different decision about the allocation of their capital resources in order to take advantage of the facility being built. As it is, the opportunity to use the new highway capacity for bus operations has now been lost without MARTA having the opportunity to strategically evaluate the services it provides in this corridor.

2. When federal SIB legislation was originally written, SIBs were intended to have the flexibility to fund either highway or transit projects. Transit advocates, fearing that highway projects would consume funds that should instead go to transit, lobbied to create separate highway and transit accounts within each SIB. Transit advocates were successful in

this effort and helped establish a system wherein funds could not be transferred from one account to the other after SIB allocations are made. The same advocates, however, have not been as vigilant in ensuring that transit accounts are funded. As a result, the rigid separation of SIB highway and transit accounts may make it more difficult, rather than easier, for transit operators to take advantage of the financing opportunities presented by SIBs.

Opportunities for New Funding Sources

Many state and local transit agencies have clearly identified and taken advantage of new funding sources to replace federal sources that have dried up. Even so, transit agencies around the country continue to confront difficult questions as they consider how to increase their public funding levels or how to pare services.

The preceding summary of transit agencies' recent funding history provides a useful backdrop for a discussion of opportunities for new funding sources. However, the ability of transit agencies to simply "backfill" for lost federal funds is not what is meant by the term "innovative financing." For the purposes of this article, innovative financing refers to increasing funds for transit by:

1. Increasing private participation in transit operations and ownership.
2. Using new financing techniques to leverage existing resources or attract new ones (some of these may also involve public–private partnerships).
3. Increasing revenue ridership.

As indicated earlier, the goal of all the efforts described below is to supplement public funds, not to replace them.

The message here is that there are many opportunities for innovative financing, but almost all involve cooperation of public entities, complex public processes and negotiation with private parties. To take advantage of any of them, transit-agency staff must be able to energetically pursue opportunities and look carefully at both benefits and potential costs and risks.

Finally, the entrepreneurial spirit advocated here should result in attempts to apply the successes experienced elsewhere to the transit setting. But it is equally important to be clear-eyed about the differences between the transit environment and other transportation environments and about the limits of innovative measures imported from other settings. Poorly planned projects are just as bad as no innovation at all.

Private Participation

Private participation in transit projects can offer many advantages. Involving private-sector entities in transit project development and operation can result in the following non-exhaustive set of benefits: increased likelihood of time-certain completion of projects; increased likelihood that projects are developed when needed, not when public funds are available; more cost-effective service delivery; and the development of innovative service delivery options. The arrangements described below all offer at least one of these types of benefits:

- ***Use of "bridge" financing by the private sector.*** The construction of the Hudson-Bergen line in northern New Jersey provides a good example of this. The use of grant anticipation notes in this project ensured that construction of the Hudson-Bergen line would begin well before Federal Transit Administration (FTA) funds were available.
- ***Linking the private sector to the development of transit-related projects.*** Ideally, the gains that result from these projects can be used to support transit investment and operations. Examples include a recent offer from a Bechtel real-estate and development partnership to build an airport transit line in Portland, Oregon, that will run through properties the partnership aims to develop; a deal with Commonwealth Atlantic Properties to build a transit station at Potomac Yards in Northern Virginia; and developer proffers of land and cash to construct park-and-ride facilities and a bus transfer center in support of the Dulles Corridor Express Bus Park-and-Ride Program in Fairfax County, Virginia.
- ***Private construction of facilities and related risk sharing.*** Because claims are a big part of traditional transit-construction financing, design-build arrangements may be able to reduce costly and time-consuming disputes. Design-build arrangements also are useful for projects that require time-certain delivery of a functional system. The recently opened extensions of Baltimore's light rail system relied on a design-build arrangement. FTA also is sponsoring a pilot program to implement design-build procedures more widely.

- ***Transfer of responsibility for some operations to private operators (paratransit).*** Taxicabs offer an example of one paratransit option, but privately operated jitneys, shuttle buses and minibuses have the potential to compete successfully for even more transit traffic.
- ***Joint development of transit facilities.*** Partnerships with telecommunications firms for right-of-way, shared facilities and rider services can enable transit agencies to share costs associated with route construction.

Innovative Financing Techniques

Most innovative finance techniques involve finding ways to reduce interest costs, increase the flexibility of repayment terms, reduce the need for up-front capital costs, share financing responsibilities with parties that traditionally have not assumed these responsibilities, or expand the range of expenditures for which local matches can be credited (typically to include expenditures that local governments would have made anyway but that also happen to support project development).

All of the innovative financing techniques described below offer these benefits, albeit to varying degrees. However, no financing technique will be able to put a project over the feasibility "hump" if revenues and costs are not in line with one another. In other words, transit system fundamentals almost always need to be addressed before innovative finance techniques will be useful.

- ***Use of state revolving loan funds, or SIBs.*** State Infrastructure Banks (SIBs) are revolving loan funds that have the potential to offer below-market interest rates and flexible repayment terms to projects that might not be built without such terms. There are only a few examples to date of projects that have used SIB financing successfully. A loan from the Florida SIB for development of Orlando's light rail system is close to closing and provides a good example. The Orlando project will be supported by future federal funds rather than by any permanent increase in funds. The Missouri Department of Transportation has proposed an intermodal facility for St. Louis, and the SIB application for Massachusetts calls for support of park-and-ride lots.
- ***Private financing/BOT structures.*** A number of examples of such projects currently exist, including the Tren Urbano, the BART extension to San Francisco International, and the Hudson-Bergen line. However, such projects are more typical abroad than in the United

States. This is because: (a) the United States enjoys more established tax-exempt debt financing for public projects; (b) other countries have a poorer public sector compared with the private sector; (c) public requirements for bidding in the United States can be relatively cumbersome; and (d) lenders like the World Bank get involved in projects in less developed countries but not in the United States.

Unlike the preceding examples, most uses of innovative financing by transit agencies appear to involve a relatively large amount of paperwork for moderate financial benefits. Examples of highly touted but frequently only moderately effective mechanisms include the following:

- ***Certificates of participation (COPs).*** COPs are issued by special-purpose entities that can purchase assets and lease them back to transit systems. An example of how these special purpose entities can offer access to savings is by obtaining discounts on bulk purchases of equipment that can then be leased to transit agencies.
- ***Cross-border leases.*** Cross-border leases involve the ownership of transit (or other) assets by entities outside U.S. borders. These assets are then leased back to the transit operator, passing through the tax advantages associated with foreign ownership. The acquisition of Baltimore's light-rail transit vehicles involved cross-border leases. Note that savings from these complex transactions rarely total more than 2 or 3 percent of capital costs.
- ***Joint development.*** Joint development involves the private development of transit-owned property or the cooperative development of property adjacent to transit facilities. Joint development projects can provide tangible benefits. For example, the Washington Metropolitan Area Transit Authority enjoys an annual revenue stream of about $5 million from leases on its joint development projects. In most existing applications, however, transit authorities have not been able to acquire significant amounts of excess real estate to develop, or are constrained in the ways that they can use such property. An important point regarding joint development projects is that they are typically not able to provide much cash up front to assist with facility construction costs.
- ***Delayed local matching.*** Delayed local matching refers to a practice (permitted by FTA) in which local project sponsors can defer their local match share of transit projects from early to later years.

- ***Toll revenue credits.*** FTA also permits toll revenues to be used to provide the local match for the nonfederal portion of a federal transit grant. This flexibility in crediting local governments for their local matches may make it easier in some locations to generate funds to support transit facility development.
- ***Benefit-assessment districts.*** Benefit-assessment districts represent the attempt to capture a portion of the value enabled by newly constructed transit facilities. Typically, a fee representing some fraction of the estimated benefit per development unit is assessed on private development that is constructed within a specified impact area around a transit station or other transit facility. Benefit-assessment districts were used in the Denver Tech Center project, but this dedicated funding source was not sufficient to meet capital and operating needs. In Los Angeles, attempts to impose special assessment districts around heavy rail stations encountered resistance in the courts for many years. These two issues—funding capability insufficient for systemwide support, and vulnerability to judicial and political challenge—will continue to characterize the funding potential and institutional environment associated with benefit-assessment districts.

Increasing Revenue Ridership

Despite the history of declining market share over the past several decades, real opportunities exist to attract new ridership—often with some combination of technology and innovative service. Successful implementation of these approaches will often require private-sector involvement. Others will require significant shifts in how transit authorities provide service. Though the focus of this article is not on innovations in transit service, these approaches are mentioned because the ability to increase transit ridership and transit efficiency has significant implications for the role innovative methods can play in the financing of transit service improvements.

The more transit agencies can close the gap between revenues and costs, the more important innovative financing methods will become to the establishment of new and improved transit operations. In other words, the smaller the difference between costs and revenues, the greater the role of financing schemes in supporting the viability of new transit projects.

Examples of methods that could potentially increase revenue ridership include the following:

- ***Use of ITS to increase ridership and reduce operating costs.*** A good example of this is New York's introduction of electronic fare cards. In this and other instances, technology offers not only a service improvement for riders but also an opportunity to develop relationships between operators and third-party information service providers through integrated, advanced passenger-information systems. Other examples include the implementation of high-tech, user-friendly fare collection, and the use of high-tech customer services such as interactive trip planning via the Internet or cable TV.
- ***Use of ITS technology to improve transit service.*** For instance, technology currently exists to allow smaller buses to provide on-call, door-to-door service in certain corridors. This has the potential to support service comparable with private automobile use at a fraction of the cost. This approach could actually involve two innovations: the implementation of ITS and the use of ITS to establish franchise agreements with small-scale operators.
- ***Joint development of transit facilities.*** Joint development is often touted for its potential to support the financing of transit facilities. However, joint development with transit-compatible land uses (mainly higher-density, mixed-use, pedestrian-friendly development) also has the potential to increase system ridership and increase revenues without raising operating costs.
- ***Improvements in transit service levels and focusing service improvements on corridors with high ridership potential.*** Adopting a customer-service orientation and adapting to changing travel patterns and purposes are crucial to increasing ridership and efficiency. These approaches could well involve cutting service to areas with low ridership and encouraging paratransit operations to serve marginally productive routes.

INNOVATIVE FINANCE

For a Bank to Finance Bridges, Roads and Sewers

A bill is pending before the New Jersey Legislature that would create the New Jersey Infrastructure Bank, a public entity through which scarce capital resources would be leveraged to increase the amount of money spent for infrastructure projects. New Jersey has received national recognition for this financing concept, which could serve as an example for the rest of the country. Several other states are seriously considering the idea.

At the core of the proposed bank is a revolving loan fund. Consumers commonly use such funds to finance durable goods like refrigerators, washing machines and cars. A variation of it could provide a practical way to finance a significant portion of the $500 billion the nation must spend over the next decade to restore its crumbling infrastructure.

A typical consumer revolving loan fund is simple. The consumer obtains a line of credit from his bank. Then he borrows against this to purchase, say, a car. He makes regular monthly payments into the fund to liquidate the loan over a period of years. Since these payments, minus interest, restore the loan fund's balance, they can be "borrowed back" by the consumer in the future to finance other purchases. The whole process can continue indefinitely, with the fund serving as a permanent source of financing for goods.

States could adopt this mechanism by establishing publicly owned infrastructure banks. These banks would make low-cost loans to state and local-government operating agencies for such capital projects as bridge reconstruction, street and highway renewal, mass-transit

---

Adapted from Joseph M. Giglio, "For a Bank to Finance Bridges, Roads, Sewers," New York Times, *October 29, 1983.*

improvements and replacement of aging water mains and sewer systems. The operating agencies would pay back the loans over 20 years or so, thus replenishing each bank's capital so it could make more loans for infrastructure projects. In effect, bank capital is recycled a number of times so that more than one dollar's worth of projects can be financed for each dollar of initial capital. In New Jersey, for every dollar of capital contributed to the bank, it is expected that at least $1.50 of capital outlays for infrastructure projects will be generated.

There are several possible sources of initial loan capital for state infrastructure banks. One is existing federal grant programs that are targeted to infrastructure projects. Another is state construction bond issues that already have been authorized or may be authorized. A third is long-term debt secured by dedicated taxes and/or fees and issued directly by the state infrastructure banks. A fourth might be new federal grant or loan programs specifically established to capitalize state infrastructure banks. Several bills have been introduced in Congress to create such programs, and the outlook for favorable action could improve if a significant number of states demonstrated serious interest in establishing infrastructure banks. None of these sources is mutually exclusive. Each state could choose from among the various funding options to provide infrastructure banks with the maximum amount of initial capital.

The infrastructure bank approach has distinct advantages over traditional ways of financing infrastructure costs. For one thing, the recycling of loan payments means that the bank's initial capital is not simply spent and never recovered (as happens with existing federal grants) or gradually returned to bondholders (standard practice for capital raised through state and local government bond issues). Instead, the recycling process keeps bank capital available for many years to provide a permanent, self-renewable source of funds for regular restoration or replacement of infrastructure facilities as they wear out. The lack of such a dedicated funding source is a major reason why the nation's infrastructure is in such poor shape.

A second advantage is that the infrastructure bank approach keeps the main responsibility for planning, funding and implementing infrastructure restoration at the state and local levels, where needs are best understood and can be handled most expeditiously. No new federal bureaucracy would be created to impose unrealistic standards and lengthy reviews, which lead to delays and higher costs.

The revolving loan fund approach to financing infrastructure should not be considered a cost-free panacea. Americans must pay fair prices for the public services and facilities they demand. Nor does it necessarily address the thorny question of how local communities—poor in capital resources yet greatly in need of infrastructure restoration—can be assisted. But it may provide a sound mechanism for dealing with these issues.

Revolving loan funds may well be a practical solution to the problem of creating a robust infrastructure capable of carrying America into the 21st century with confidence and strength.

THE POWER OF TECHNOLOGY

Finance and the Visible Hand of Technology

Finance is about more than just money. This is not an easy sentence to say with a straight face, nor one that many people necessarily would believe. Obviously, money itself is very important, since how much we have determines how much we can build and how well we can manage the assets we have.

The general consensus is that money will be harder to find for the upcoming reauthorization of the Transportation Equity Act (TEA-3) than it was for the original legislation (TEA-21), a result of improving fuel economy, slower economic growth, higher oil prices, a tighter federal budget and perhaps some resentment over the success achieved six years ago. This implies we need ideas that involve more than changes to our current financial toolbox. Such changes, in turn, may open the door for approaches that go beyond the traditional broad-based user fees, with or without some leveraging. My discussion today will not solve any immediate problem, but it could help with the next big piece of legislation—TEA-4.

Background

How we raise money affects what we do and how well we do it. Here I am not just talking about TIFIA versus federal apportionments or one flavor of GARVEE bonds versus another. Rather, the issue relates to the fundamental economic linkages between prices and service quality. Put

---

Adapted from Joseph M. Giglio and Daniel J. McCarthy, "Finance and the Visible Hand of Technology," in Transportation Finance: Meeting the Funding Challenge Today, Shaping Policies for Tomorrow: Report of the Committee for the Third National Conference on Transportation Finance, Chicago, Illinois, October 27-30, 2002, *80–84 (Washington, D.C.: Transportation Research Board, 2005).*

another way, in the best of all worlds we should be able to generate funds in ways that encourage better service and operating efficiency as well as that provide an adequate level of funds.

This sounds a lot like what a successful private-sector business does—simply provide a useful service, control costs, and generate enough money to cover capital and operating costs as well as a return for future investment. Indeed, finance is about more than just money; finance is about prices. In the private sector, prices reflect the quality of the product sold or the value of the service provided. That is, the amount of revenue a firm books relates directly to the quality of the service that it provides relative to that of its competition.

Feedback between quality and the amount of business a firm receives can be rapid, particularly in businesses that deal directly with the retail customer. This is often very personal. We all have been asked by our waiter or waitress about the quality of our meal—and may even have had an entrée dropped from the bill if we expressed dissatisfaction. How often has the Secretary of Transportation or District Engineer asked us if we were satisfied with our daily commute, let alone offer a refund as a sign of their commitment to improve?

This process is part of what Adam Smith termed "the invisible hand" that balances supply and demand and thus shapes the economy. While highway finance has long relied on a set of broad user fees or benefit taxes, these reflect long-term values rather than near-term service quality. As long as the level of highway revenues bears at best an indirect linkage to the quality of service received by the public, we will be forced to work with a very "visible hand." Intelligent Transportation Systems (ITS) and telematics technologies may offer a way to make the idea of a "visible hand" practical.

In sum, I believe there is an interaction among finance, technology, customer service and the management of Departments of Transportation. If we look at just finance alone, we risk continued frustration—and a transportation system that fails to meet our economic and social needs. If we work to improve management alone, we miss using one of the more effective tools for efficiency and customer satisfaction.

To set the stage for this argument, I will address two trends that are broader than finance itself. First is the move by Secretaries of Transportation and many other transport administrators to emphasize their role as managers of a business. Second are the changes in technology that make it possible to measure highway performance directly.

Management

To date, the most imaginative DOT leaders think and act like they are running a business—indeed, most state Secretaries of Transportation now call themselves CEOs rather than CAOs or Chief Administrative Officers. This has generated interest in a host of new management techniques and approaches. For example, asset management is now a hot topic, albeit one that remains dominated by engineering-based techniques and measures, rather than financial measures and economic efficiency.

Asset management offers promise, but few systems have yet to be implemented that examine assets the way a private enterprise would. One reason is simply that our surface-transportation systems do not operate in a free market. Our ability to generate a real breakthrough in this regard could take one of two paths:

1. ***Direct competition.*** To be effective, this needs to be more than transit versus highways since it is tough to have competition when one entity has 95 percent of the market and its competitor five percent. Thus, despite the ideological appeal of public-private partnerships, we are unlikely to have more than one DOT per state.
2. ***True measures of performance.*** If real competition is unlikely, then at least we can try to improve our measurement of service. Once we measure the performance of the highway system on a consistent and routine basis, then we can start to develop a pricing system that sends market signals to travelers and providers alike.

Because there is no true competition for local highways and transit, there is no incentive to develop true measures of customer satisfaction or system performance. How do we do this without the invisible hand of the marketplace? Perhaps we can use technology to create a smarter, visible hand. This will never replace the elegance and effectiveness of Adam Smith's "invisible hand," but a good second best is much better than what we have now.

Transportation depends on assets, but this emphasis ignores the routine economic and social values transportation provides. For example, while volume-to-capacity ratios and levels of service provide one measure of congestion, the consumer of transportation services probably relies on a much more basic measure: Did I get where I want to go on time? Or more broadly, what fraction of peak-hour trips arrived on time this morning? What fraction of supply-chain shipments arrived on time?

This is similar to measures used by the airline industry today. Again, this requires a direct and personal measure of highway performance—something that a vehicle-oriented technology might provide.

TECHNOLOGY

Technology offers a possible way to measure direct performance. Certainly one dream of the ITS community has been roadway sensors that make it possible to measure traffic speeds on all roadways, all the time. Such information could then be used both to manage the system better as well as to provide information to the traveling public. In reality, over the past dozen years, the pace of deployment has been glacial, with only 22 percent of urban expressways having sensors of any kind by 1999 and less than 10 percent of major arterials.[1] In recent years the rate of growth has rocketed to 3 percent a year—meaning the problem will not be solved in my lifetime.

As Phil Tarnoff from the University of Maryland points out,[2] advances have been made that include:

- Sophisticated side-fire radar detectors capable of providing traffic flow data by lane are currently in use.
- Video imaging that can now provide reliable measurements of traffic speeds, counts and queues is also in use.
- Cellular geolocation technology with the capability of tracking individual vehicles in order to measure their speeds and travel times has been successfully demonstrated.
- Tracking of vehicles with toll tags to measure speeds and travel times is currently in operation.
- Tracking of vehicles through the use of license plate readers that measure speeds and travel times is also currently in use.
- Systems of instrumented probe vehicles using GPS for positioning and other sensors to measure travel speed, weather and pavement conditions are also under development.

Frustration with the slow pace of deployment despite the range of technical solutions has led to some new approaches. The U.S. Department of Transportation's INFOstructure plan, for example, calls for a nationwide network of traffic sensors and video monitors. This is ambitious and also offers benefits for Homeland Security. On the other hand, the cost is very high ($5 billion has been mentioned) and it faces a problematic future in our new world of growing budget deficits. It also represents a continuation of existing technologies, just with a fixed schedule and more funds to back them.

Technology, however, can also develop along non-linear or divergent paths. These types of changes are hard to foresee, in part because many fail or remain dormant until the time is right. When they do occur, they can create a chain reaction of changes well beyond the immediate area of focus. Examples abound in our daily lives, with the most obvious coming from the Internet, wireless communication and the personal computer. Examples from previous generations include the automobile, jet aircraft, the U.S. Interstate Highway System and so on. While progress continues, transportation technology has yet to see such a breakthrough.

Within the world of traffic sensors, we may be on the verge of a non-linear shift. This revolution will be built on vehicle-based sensors, rather than those that report from specific points along the highway infrastructure. Floating car data and vehicle probes appear to be more powerful and cost-effective options than fixed, infrastructure-based sensors.

Vehicles offer several advantages:

- They provide a direct measure of performance as seen by the roadway customer, rather than measures inferred from volume-capacity ratios.
- They can do this across all roads in all parts of the country at the same time.
- They can provide measures that are consistent from road to road and from city to city.
- They provide geographic detail that can be reshaped to meet a variety of specific public and private purposes.

Such data, in turn, will support most traditional traffic-management activities. Because of the direct link with roadway performance, they also provide the foundation for a profound shift in transportation management. This has implications for activities from day-to-day management to financing to emergency support. The same data provide the long-sought information needed by the traveler-information portion of the telematics industry. It may even be possible for more than one of these firms to make money!

Vehicle-based sensors are well along the development path. Floating car data systems exist on small scales in Europe. OnStar in the United States has units in some three million vehicles. These represent more than one percent of the nation's fleet—this should be enough to estimate travel speeds by roadway link. Of course, a number of communication and financial issues need to be resolved before this becomes commercial reality. At the same time, some 600,000 commercial vehicles already have tracking equipment as part of the fleet management industry, with Qualcomm the leader. Other systems propose the use of cell phones as data probes to provide location, speed and acceleration information. While well behind schedule, progress is underway to enforce the FCC's E-911 mandate to convert the nation's 100 million cell phones into probes.

As one example, the British firm ITIS Holdings has deployed a floating vehicle system that converts high-mileage vehicles into probes and provides regular reports on more than 8,000 miles of motorway and major arterials in the U.K. Their commercial customers include the British AA, BMW and OnStar. Other than this, no vehicle-based system has been deployed on a significant scale. All have one or more problems to resolve, mostly finance related.

Regardless, the time is right for a new way to collect traffic data. The technology exists and several commercial enterprises have begun to deploy their networks. Vehicle-based systems will support activities and businesses well beyond traditional traffic problems. How might the public sector play a role in this movement, and what implications might this have for how we manage and fund transportation?

IMPLICATIONS

If through some bizarre twist of fate, I were named benevolent transportation czar, I would want to manage and fund my transportation system a bit differently. I would take a lesson from the best of the private firms and set a one key goal: Why not work so that all important trips are completed on time—or at least better than the airline industry. Such an objective would have implications for a very different transportation system, both in how it is managed and how it is funded:

- Part of this new transport system would involve having information on system reliability that could be communicated to travelers in advance so that they could manage their own travel. I am not naive enough to believe that this would cause people to shift modes. However, this information could enable them to change the time they travel, shift routes to take full advantage of available capacity, and call ahead to minimize disruption.

- Part of this would mean that I would manage my system to a pre-set performance standard—requiring, for example, average speeds of better than 50 miles per hour, all routine maintenance and construction work done in off-peak hours, all incidents cleared in so many minutes, and so on.
- Part would involve a more personal relationship with the traveler. I would even go so far as to provide rebates if performance standards were not met. Again, technology in the form of transponders offers a direct way to implement a "money-back guarantee."
- Part would also involve knowing the relative value of a timely trip completion (just-in-time inventories provide a direct example for freight).
- Part would also involve a direct cooperation with certain major transportation customers, ranging from sporting events to job locations to individual industries. This knowledge would provide a source of money (as shown by the use of variable tolls on S.R. 91 and the San Diego HOT lane to ensure a given quality level). It would also require a direct cooperation between the transportation provider and its customers
- Part would involve a very different set of internal standards for district engineers. For example, knowing that it was possible to measure the average speed for the morning or afternoon commute would change the incentives for the district engineers. The British highway authority already does this when they outsource maintenance work, with part of the compensation dependent on the lack of congestion as measured by the amount of time that design speeds are met.

In closing, let me return to the finance question. Today, highway travelers in the United States pay a low average rate—only pennies per mile traveled. This is much less than the 35 cents per mile that it costs to own and operate their vehicles. Today, financing is based on the long-term average cost of highways, with a correspondingly average quality of service. The ability to provide a high-quality service begins with the ability to measure performance, but it allows a price that more closely reflects the value of the completed trip.

In sum, the technology to measure highway performance on a link-by-link basis also opens the door to a host of financial and performance initiatives. This creates the opportunity to unleash some of the creativity and management techniques that the "invisible hand" of the marketplace stimulates in other industries.

The gains from this could occur along several dimensions. While only one of these changes has a direct link to increased funds, each should improve the financial health of the transportation agency. Examples include:

- ***Improved customer satisfaction.*** Most travelers take traffic for granted as something imposed on their daily lives, much as the weather. Direct measurement of highway performance will permit a host of changes that will show the DOT cares—money-back guarantees in case service is poor.
- ***Improved day-to-day management.*** Most DOTs rely on important, but indirect measures. A more direct set of measures—such as average speeds during the morning commute—will result in improvements once it is possible to measure performance directly,
- ***Increased revenues.*** Higher quality service is not free. The technical ability to charge travelers more when the service is better (the flip side of a money-back guarantee) will also generate more resources. This will be done independently of automobile fuel economy and the motor-fuel tax in general.
- ***A more efficient economy.*** The combination of performance measures related to travel times and financial incentives to encourage improvement should also improve economic productivity.

NOTES

1. Phillip J. Tarnoff, "Getting to the INFOstructure," prepared for Transportation Research Board Roadway INFOstructure Conference, August 2002.

2. Ibid, p. 7.

THE POWER OF TECHNOLOGY

The Impact of Technology on Roadway Operations and Financing

Recent advances in electronic technology have opened the door to nothing less than a revolution in the way we operate and finance our roadway systems. For the first time, we're able to price roadway use with the same flexibility as private business firms have traditionally enjoyed in pricing access to their services.

The Movie House Analogy

To appreciate what this means, let's assume for a moment that we own a five-screen movie house in a suburban shopping mall. The whole basis for how we operate our movie house is that our customers must first buy tickets at the box office before they can enter any one of our five theaters to enjoy a movie. Because we want to generate as much revenue as possible, we try to adjust ticket prices so that we fill our seats with customers who buy tickets at the highest prices they're willing to pay. So we will obviously charge higher prices for evening tickets and lower prices for afternoon tickets to reflect different levels of demand at different times of the day. We will also increase prices for both evening and afternoon tickets on Saturdays and Sundays, the most popular movie-going days during the average week. And we'll do the same thing for holidays.

These actions reflect the fact that at any given time of day (and day of week), a certain number of people may be willing to see movies in our five-screen movie house. But the size of this number depends in part on

---

Adapted from Joseph M. Giglio, "The Impact of Technology on Roadway Operations and Financing," (paper delivered to the Transportation Research Board National Conference on Transportation Finance, August 2000).

how we price these tickets. If we charge too little on Saturday night when demand is high, we can keep our theaters full. But we may end up turning away potential customers willing to pay higher ticket prices than those filling our seats. So we've lost a revenue opportunity. If we charge too much on weekday afternoons when demand is lower, we're likely to have a lot of empty seats. But we might have been able to fill at least some of seats if our ticket prices were lower. So we've lost another revenue opportunity.

In other words, the whole process of deciding ticket prices is something of a juggling act. As with all such juggling acts, the key to success is staying alert to changes in circumstances and moving quickly to exploit them. All the while, we must keep in mind that our goal is to fill as many seats as possible with customers who buy tickets at the highest price they're willing to pay.

Of course, this "highest acceptable price" often depends on the popularity of the movie being shown. All movies aren't equally popular. So our ticket pricing strategies should take this into account.

For example, each of our five theaters shows a different movie on any given Saturday night. If one of these movies is significantly more popular than the others, the length of the line outside the theater may discourage some potential customers from waiting and cause them to go elsewhere. Meanwhile, the other four theaters showing less popular movies may have a number of empty seats. So why not price access to each theater in a way that reflects the relative popularity of the movie it's showing? We can raise the ticket price for the most popular movie and lower the price for the others. If we do this cleverly, we'll be able to equalize demand for access to each of our five theaters.

Some of the people standing in line for the most popular movie may be lured by lower ticket prices to the other four movies. And those left waiting in the now-shorter line are less likely to take their business elsewhere. Under the best circumstances (and best balance of prices), all five of our theaters will be full and no one will be waiting in line to see the most popular movie. This idea of **PRICE RATIONING** is a basic feature of free-market economies. It's regarded as the most efficient way to allocate goods and services—like access to movie seats—that can't be supplied in unlimited quantities.

The alternative to price rationing is **TIME RATIONING**. This is what happens when customers have to wait in line to see a popular movie. Time rationing was a standard practice in the former Soviet Union and other Eastern European countries. Although prices for consumer goods and services were kept low enough for everyone to afford, you often had to wait in line for long periods to make purchases.

This practice favors customers who place a low value on their time. But it penalizes those who are short of time and who might be willing to pay a premium to avoid waiting in line. In our movie house, we can offer customers this opportunity by charging the highest admission price for the most popular movie.

Interestingly, we Americans have traditionally used the "Leninist" practice of time rationing to control access to our roadway systems. With few exceptions, we don't charge drivers to travel on them. But even on toll bridges and highways, drivers must be willing to endure slow trips in stop-and-go traffic if they want to use roadways during periods of high demand. We don't offer them the option of paying a premium price to avoid slow trips.

That's the whole point behind the movie-house analogy. The manager of a movie house has always been able to price tickets according to demand—and give customers the option of avoiding long waits in line by paying higher ticket prices. The technology for doing this happens to be very simple. But until recently, we've lacked the advanced technology needed to price roadway access the way we price access to movie seats.

Applying Movie Ticket Pricing to Roadways

All this is finally changing. Recent advances in electronic technology have greatly simplified the process of collecting tolls, monitoring traffic volumes and sorting out different types of motor vehicles. This gives us the movie-house manager's flexibility in how we charge motorists for access to roadways, and not just on bridges or limited-access highways. Today's technology allows us to charge for access to every kind of roadway—even local streets.

This paves the way for a total revolution in roadway finance and operating efficiency. We can now charge variable toll rates based on distance, the time of day or week traveled, the roadway or lanes used, and the size and type of the vehicle driven. The later option can open the way to more effective and less costly enforcement of posted regulations prohibiting the use of certain local streets by trucks and other large commercial vehicles. A prohibited vehicle automatically gets charged a rate so high that the regulations become largely self-enforcing. We can use the same principle to enforce other traffic regulations like speed limits more effectively and less expensively.

The new technology that makes these and other benefits possible is simple, reliable, and surprisingly inexpensive:

- Simple transponders placed on vehicle windshields can transmit the vehicle owner's unique ID number to electronic monitors along a roadway. By reading this number when the vehicle enters and leaves the roadway, we can compute the distance it travels and charge the vehicle owner's account accordingly. At the appropriate per-mile rate for that type of vehicle. On that particular roadway. During that particular time of day. When the roadway is experiencing that particular traffic volume.
- Wire loops beneath roadway pavement can continuously measure average traffic speeds on each lane, allowing us to automatically adjust the per-mile rate we charge motorists. Raising rates when the speed falls too low will deter some motorists from using the roadway, thereby reducing traffic volume to a level that permits the higher average speed we desire.
- Pattern-recognition software in closed-circuit-television monitoring systems can count the number vehicles in roadway lanes. It can even give us continuous volume readings broken down by each particular type of vehicle.
- Variable message signs along each highway can provide motorists with information about the current price rate for using the highway, the current average traffic speed, problems up ahead and even conditions on parallel roadways. And it won't be long before these signs can effectively be "moved inside the vehicle" to provide motorists with price and speed data for every roadway in the system, plus on-demand travel information about optimal routes and other useful services.

Intelligent use of this technology can enable us to narrow the often-considerable gap between a roadway system's THEORETICAL CAPACITY and its FUNCTIONAL CAPACITY. By using different price levels to shift trips to less crowded roadways and lower demand periods, we can make more efficient use of our roadways. It all comes down to the classic economic principle of using price to control the demand for scarce resources. In the process, we're able to provide better service for our customers.

Yes, customers. Not motorists or travelers or taxpayers, but customers. Just like the people who buy movie tickets. This distinction is very important. If we're going to successfully run our roadway systems in a more business-like way and charge fair prices for using them, we must develop a business manager's habit of regarding those who use roadways as customers first and foremost.

What exactly is a customer? Someone who's a willing buyer of what we have to sell at the particular price we're charging. What makes this someone a willing buyer? A personal judgment about whether the value to him or her of greater than the price we're charging.

Management guru Peter Drucker once pointed out that the most important goal for every business firm is to create CUSTOMERS—with profit simply being one of the costs we have to cover to stay in business. But Drucker also insisted that the goal of creating customers is just as important for public agencies as it is for business firms.

How do we create customers? By heightening the perception that what we're selling is worth more than the price we're charging. We do this by improving the quality of what we're selling, or by reducing its price, or by doing some of both.

Suppose a driver can use two different highways to reach his destination. One highway charges a relatively high toll rate per mile but promises an average speed of 55 mph. The other highway charges nothing—but it's choked with slow, stop-and-go traffic moving at less than 20 mph. If the driver is on his way to an important business meeting and can't afford to be late, he may decide that the value of the time he'll save by using the high-toll highway is greater than the price he must pay. But if he's simply making a trip to a shopping mall to buy gardening tools, he may opt to use the toll-free highway and put up with the additional travel time.

In other words, roadway pricing lets us create value for drivers by offering them shorter travel times for their high-priority trips. Who determines the priority of trips? The drivers themselves. All drivers make personal judgments about what trips are the most important and how much they are willing to pay to reach their destinations faster. In exchange, we can offer the same kind of money-back guarantee that a movie-house manager offers if the projector breaks down or the sound system goes dead.

Pricing access to roadways in the same way we price access to movie seats enables us to operate roadway systems more efficiently. Just like a smart movie-house manager, we can use the price mechanism to distribute travel demand at any given time in a rational manner—so that traffic volumes on each roadway reflect its capacity. If we do this intelligently, we'll end up minimizing the aggregate trip times for all drivers. Shorter trip times mean better service for travel customers, and allow our roadway systems to accommodate more trips per hour.

THE REVENUE ADVANTAGE

Roadway pricing also produces revenue. Lots of revenue, if we apply this concept to an entire metropolitan roadway system. So much revenue that we can abandon forever the crude, inefficient mechanisms upon which we've had to rely in order to fund roadway operations and construction. We can abolish sales taxes on motor-vehicle fuel and tires, eliminate high registration fees based on a vehicle's weight or purchase price, tear down all those toll booths that slow traffic to a crawl during high demand periods, and end the practice of using general tax revenues to support our roadways.

Instead, roadway systems can become entirely self-supporting. Just like movie houses, they can be paid for by the people who actually use them—according to how much they use them, when they use them and the type of vehicle they are driving. Among other important benefits, making our roadway systems self-supporting can enable us to resume the too-long deferred task of expanding their capacity to accommodate the new travel demands generated by rising economic activity.

The federal government has been heavily promoting the idea that "we can't build our way out of congestion." It does this to discourage state transportation departments from planning expensive new roadway projects and then coming to beg Washington for federal grants to help fund their construction. The feds would rather pay off some of the government's debt than provide the nation with the new transportation and other infrastructure capacity it needs. This is like the homeowner who uses his latest raise to accelerate his mortgage payments rather than replacing his leaky roof or installing a more efficient heating system.

We can't build our way out of congestion? Of course we can. And we must if the nation is to prosper in the future. But there is no way to do this under today's inefficient and ineffective mechanisms for funding roadways.

Comprehensive roadway pricing gives us a sensible way to fund this badly needed new construction. The revenue streams generated by charging motorists fair prices for roadway use can support new transportation bond issues that spread capital costs over future years, reflecting the many years of benefits new roadway projects can produce. Benefits that are paid for by those who actually use the roadways.

Equally important, the realistic expectation of such revenue streams also enables us to bring together government and business in formal partnerships that exploit both public and private resources more effectively. We can do this by making greater use of a mechanism called **PROJECT FINANCING.**

Project Financing is a low-key term for a high-key approach to financing the construction of capital facilities. The private sector has used it successfully to build electric power plants, oil pipelines, even Euro-Disneyland outside Paris. But we can also use it to build new highways, bridges, transit lines and other public infrastructure facilities.

A more descriptive term for this approach is "asset-based financial engineering." In other words, the ability to raise construction funds depends entirely on the anticipated financial strength of the new capital asset that will be created.

This is a marked departure from traditional capital financing, which depends on the real or perceived financial strength of the private firm or government unit that will own and operate the asset. That's why asset-based financing is widely used by commercial real-estate developers to finance the construction of office buildings, shopping malls and large residential complexes.

As we'll see, focusing on the financial strength of the capital asset itself can make a world of difference when it comes to raising the funds we need to expand the capacity of our roadway systems.

HOW PROJECT FINANCING WORKS

Let's consider one example of how a partnership between government and private firms can use Project Financing to build a major new transportation facility. The details are important because they illustrate the pervasive nature of the benefits new technology offers for operating and financing roadways systems more rationally.

We'll call the transportation facility in question **Metro West Turnpike.** It's a new expressway that adds badly needed transportation capacity to a major travel corridor in a metropolitan region whose existing highways are heavily congested during much of the day. The expressway is designed to support itself with motor vehicle user charges, plus certain other income.

This particular travel corridor is defined by a classic 1950s urban expressway whose capacity has long since been overwhelmed by growing travel demand. But because the expressway runs through a number of densely developed areas, simply widening its right-of-way to add more lanes isn't a feasible option. The only practical solution is to build what amounts to a second expressway on an elevated structure over the existing expressway. The state's Department of Transportation previously handled all new highway construction in the traditional manner. This includes relying on grants from the federal government's Highway Trust Fund to cover most of the capital cost of building new highways.

But such grants have become increasingly difficult to come by now that Uncle Sam has put on his Uncle Scrooge hat. The DOT finally comes to terms with the idea of building the new expressway as a self-supporting toll road. This opens the way to funding construction costs by issuing bonds against future toll revenues, and eliminates any need for federal grants—not to mention the lengthy grant application process.

Now the state DOT takes the next step and establishes a public/private partnership to build and operate the road. This offers two additional advantages. The partnership can use the Project Financing approach to raise construction capital from a broader spectrum of sources than is available for traditional public financing. And some of this capital can take the form of equity contributions from the partners themselves. These partners may include certain business firms that have vested interests in seeing the highway built.

All this leads to a financial engineering structure whose key element is Metro West Turnpike LP (MWT for short). MWT is the legal entity that finances, builds, owns and operates the expressway. The "LP" at the end of its name means that MWT is a "limited partnership" rather than an ordinary private corporation. Being a limited partnership has certain advantages that will become apparent as we explore its structure.

MWT follows a three-stage process to build the expressway and get it operating.

Stage One: Construction

MWT uses the equity capital contributed by its partners, plus a construction loan from a consortium of commercial banks, to fund the expressway's construction and initial financing costs.

Stage Two: Refinancing

At the end of the construction period, MWT sells long-term mortgage bonds that are secured by a lien on its assets (which means the expressway itself). The interest and principal payments on these bonds will be covered by the income that the expressway generates from tolls and other sources. MWT uses the proceeds of this bond sale to pay off its construction loan.

In other words, MWT converts its short-term bank debt into long-term mortgage debt once the expressway is ready to begin operating. This is the same two-stage financing approach commonly used by private real-estate developers to fund the construction of new office buildings and other projects.

Stage Three: Operations

Once the expressway opens for business, MWT uses the income it generates to fund its annual costs. These include:

- The expressway's normal operating and maintenance expenses.
- Interest and principal payments to MWT's bondholders.
- Maintenance of suitable levels of capital reserves, which MWT funds with the portion of its operating revenues that is allocated to annual depreciation of its capital plant and equipment.
- Annual dividends payments to MWT's partners.

That's the whole story in a nutshell. But we need to look more closely at some of the details in order to understand the important subtleties of Project Financing.

MWT's Partners

Some of MWT's partners are simply clients of securities brokers and other private investment firms that are always looking for new investment opportunities. The projected level of MWT's dividends makes it an attractive investment, especially since these dividends will be partially sheltered from income tax liability (more about this in a moment).

But MWT's partners also include various entities that have some kind of vested interest in seeing the turnpike built and are willing to contribute equity capital to help accomplish this.

These partners include:

- The state's Department of Transportation, which conceived and planned the expressway and established MWT.
- Three suburban governments whose jurisdictions are served by the expressway and various construction firms that receive contracts to build the expressway. (Since MWT is a private-sector entity, it doesn't have to follow the standard, arms-length, competitive bidding process required for government agencies.)
- Firms that supply materials and equipment used in building the expressway.
- The engineering firm that receives the contract to design the expressway and manage its construction.
- An existing public toll-bridge authority, to which MWT awards a contract to manage the expressway's daily operations and maintenance.
- Several telecommunications and electric power companies wishing to rent right-of-way space along the expressway to install new transmission cables.
- Large trucking firms hoping to benefit from the additional transportation capacity that the expressway provides in a high-demand travel corridor.
- Distributors of petroleum products who believe that the expressway will help induce more daily motor-vehicle-miles-of-travel in the region, which will increase sales of gasoline and diesel fuel.

Tax Subsidies

Built into the federal, state and local income tax codes are certain cost-saving deductions available to taxable private enterprises but not to individuals or government agencies. These deductions are commonly known as tax subsidies. The two most important are:

- ***Deductions for annual interest payments on outstanding debt.*** Unlike the interest payments an individual makes on his personal credit card debt, all interest payments made by private business firms are fully deductible in determining their taxable income. In MWT's case, the effect is to reduce its out-of-pocket interest costs to levels that can be very competitive with interest costs on tax-exempt bonds that the state government might otherwise have had to issue to fund the expressway's construction cost.

- ***Deductions for annual depreciation of capital plant and equipment.*** The tax codes allow MWT to depreciate these assets more rapidly for tax purposes than it does for regular accounting purposes. Since depreciation doesn't involve any actual cash outlays (as do such deductible costs as employee salaries and payments to outside suppliers), this amounts to a deduction for costs MWT didn't have to cover by writing checks during the tax year. Another way to look at tax depreciation is as a process for recovering invested capital more rapidly.

As a private, profit-seeking enterprise, MWT has income-tax liability. But since it is also a limited partnership, this liability is passed along to its partners with the dividends they receive. However, the tax deductions for interest and depreciation are also passed along to the partners. So a significant portion of their dividend income is tax-free.

Motor Vehicle Customers

A crucial component of the MWT enterprise is the drivers who pay tolls to use the expressway. Electronic toll-collection technology enables MWT to charge drivers on a per-mile basis rather than a traditional flat fee. The same technology also lets MWT vary the toll rate throughout the day in order to reflect actual demand.

The state's transportation planners did something very interesting when they conceived the expressway as a toll road. Instead of blindly seeking to maximize toll revenue by maximizing traffic volume, they decided to have the expressway cater to drivers who place a high value on their time. These drivers (most of whom are making business-related trips) are given the option of paying for faster trips on the expressway because its traffic volume is deliberately kept low enough to assure a relative high average speed. Traffic volume is regulated by the toll rate per mile, which fluctuates with demand and effectively prices many drivers off the expressway during high-demand periods. These drivers have the option of using the old expressway, which remains toll free.

Special television cameras and loop detectors continually monitor average traffic speeds on the new expressway. When its traffic volume rises to a point where this speed falls below 45 mph, the toll collection system's computer boosts the toll rate—and keeps boosting it to reduce traffic volume until the expressway's average speed moves above 45 mph. But during periods of less demand when the average speed rises above 55 mph, the system's computer reverses this process. It lowers the toll rate in incremental steps to encourage sufficient additional traffic volume for average speed to fall back below 55 mph. Drivers approaching each of the expressway's entrance ramps see the current average speed and toll rate clearly displayed on variable message signs.

In other words, from a purely business perspective, the expressway maximizes its toll revenue by pricing transportation access as high as travel demand warrants at different times throughout the day. Just as a movie-house manager prices his tickets according to demand.

Equally important, the expressway plays a rational transportation role by deliberately restricting itself to serving only "high priority" trips during periods when travel demand in the corridor is heavy. Drivers themselves determine these trip priorities. They make their own individual judgments about whether the travel time saved by using the expressway at any particular time is worth the posted price. If they decide that the price is too high, they can use the toll-free lanes on the old expressway and put up with slower trips because of heavier traffic volume.

Financing Costs

It costs money to raise capital funds. That's why investment banking is one of the most highly paid professions in the United States. MWT's financing costs include the various underwriting fees paid to place its limited partnership shares through brokerage firms and to issue its long-term bonds, plus the origination fees paid to the consortium of banks that provide the construction loan. Also included is the capitalized debt interest during the construction period and the various reserves that MWT had to establish to underpin its bonds.

Other Income Sources

While toll revenues provide most of its income, MWT aggressively exploits the expressway's potential to generate other kinds of income, including:

- Annual payments from electric power and telecommunications companies that lease right-of-way space along the expressway to install their transmission lines.

- Sales of commercial time to local advertisers on MWT's radio station, whose main purpose is to provide expressway customers with current information about traffic conditions.
- Sales of advertising billboard space on the expressway's variable message signs, which are located at each entrance ramp and at various points along its lanes. (As a "private roadway" that was built without any federal funds, the expressway isn't subject to federal regulations covering billboards.)
- Fees paid by local towing and emergency auto-repair companies for the right to provide their services to the drivers of vehicles that become disabled on the expressway.
- Interest earnings on MWT's overnight cash balances.

Annual Expenses

MWT's annual expenses will include:

- ***Operating and maintenance costs.*** They include the salaries of MWT's employees, payments to outside suppliers, and fees paid to entities like the local toll-bridge authority to which MWT contracts certain expressway management responsibilities.
- ***Depreciation.*** As noted earlier, MWT (which is a private-sector entity that is therefore subject to the accounting standards governing all private firms) allocates a portion of each year's operating revenue to cover annual depreciation of the expressway's capital plant and equipment. This is regarded as a regular operating cost, just like employee salaries, and must therefore be covered by operating income. Funds allocated to depreciation are then transferred to MWT's capital reserve account.
- ***Other costs.*** These include the usual variety of miscellaneous costs incurred by all business enterprises.

MWT's Bondholders

Most of MWT's bondholders are life-insurance companies, private pension funds and other large institutional investors.

Since a portion of MWT's bonds mature each year, these investors can buy bonds that promise the repayment of their invested funds on schedules that meet their estimated cash needs in the future. Meanwhile, they earn attractive interest income on their funds.

Why Project Financing Works

The underlying premise of Project Financing involves a process of deliberately "commercializing" the services that a capital facility is built to provide.

This means these services must be perceived as necessary or desirable by enough customers who are willing to pay a sufficiently high price for them. In MWT's case, most of these customers are drivers seeking to save travel time. Other potential customers include various businesses that believe they can profit by paying MWT for the right to use its right-of-way space, to advertise on its radio station and variable message signs, or to pay for the right to provide emergency road services to expressway travelers.

When this perception of necessity and desirability is achieved, the capital facility is able to generate a sufficiently large and reliable stream of business-type revenue. For the purposes of discussion, let's say MWT generated $925 million in business-type revenue by selling its services to drivers and other customers during the first five years of the expressway's operation.

A "sufficiently large" revenue stream is one that can comfortably cover 100 percent of the facility's annual economic costs. Not the least of these costs is an adequate return on the equity and debt capital that the facility's owners and bondholders invested to fund its capital costs. MWT paid $320 million in dividends to its partners during the first five years of the expressway's operation, and made interest payments of $90 million to its bondholders during this period. These payments were in addition to MWT's $450 million in operating expenses, $120 million in depreciation, $40 million to retire maturing bonds, and $25 million in other expenses.

How large a return is "adequate" depends in part on the reliability of the revenue stream. Low levels of reliability mean greater risk for the investors who supply the capital to build the facility. Greater risk must be offset by a higher return—and vice versa. During its first five operating years, 86 percent of MWT's operating revenue came from expressway tolls that rose steadily because increasing travel demand in the corridor results in both growing patronage and growing average daily toll rates.

These Stage Three cash flows underpin a public–private partnership's ability to raise capital funds for construction by using the Project Financing approach. The basic principle is that reliable projections of a capital facility's future operating cash flows will determine how much capital the partnership can raise and what sources it can tap.

In MWT's case, Stage One begins by raising $160 million in the form of equity contributions from its partners. With this equity capital in hand, MWT obtains a $400 million construction loan from a consortium of commercial banks that have considerable experience making similar loans to private real-estate developers. The basis for this loan is the assumption that MWT will pay it off at the end of the construction period by issuing long-term bonds, just as real-estate developers pay off their construction loans by taking out long-term mortgages on the commercial buildings they develop.

MWT now has $560 million in capital funds. After deducting financing costs, this is enough to cover most of the expressway's construction costs. Towards the end of the construction period, MWT makes its first foray into the public debt market by issuing $225 million in bonds. After deducting financing costs, this is enough to complete the expressway and ready it for operations.

When the turnpike is finished, MWT proceeds to Stage Two. This involves refinancing its construction loan. MWT does this by issuing a second round of long-term bonds to raise an additional $525 million. After setting aside $125 million to establish the necessary debt service reserve and cover other financing costs, MWT uses the remaining $400 million to pay off its construction loan.

Upon completion of Stage Two, MWT's capital totals $910 million. This consists of $160 million in equity, 17.6 percent of the total capital, and $750 million in long-term debt, 82.4 percent of the total capital. This debt is scheduled to mature in back-loaded annual installments over the 25 years of the expressway's expected useful operating life before it will require major capital reconstruction.

Nearly all of the capital is made up of private-sector funds that are supported entirely by MWT's operating income from commercializing its services to drivers and other customers. The only public-sector funds in this mix are the equity contributions made by the state and local governments. These are also supported by MWT's commercial income, not by tax revenues.

CONTRASTS WITH TRADITIONAL PUBLIC FINANCING

The typical "all-public" approach to building and operating highways and other infrastructure facilities normally avoids any attempt to commercialize the services they provide. Such facilities are supported entirely by the larger community of taxpayers, whose tax dollars provide the funds needed to cover operating and maintenance costs, payments of interest and principal on construction bonds issued by state and local governments, and federal grants.

The standard justification for the all-public approach is that infrastructure facilities provide a complex structure of direct and indirect benefits to the community as a whole, not just to the users of these facilities. These benefits, the thinking goes, lead to higher levels of economic activity in the community, which translate into higher incomes for individuals and higher profits for business firms. Furthermore, the dollar value of these benefits to each taxpayer is assumed to be reflected by the size of an individual's income and a business firm's profits. By taxing these incomes and profits on a more or less proportional basis, government is therefore charging individuals and business firms what amounts to "a fair annual fee" for the benefits they receive.

Both the all-public approach and the public-private partnership approach ultimately rely on a REVENUE BASE, which is composed of the individuals and business firms in the community.

- In the all-public approach, these individuals and business firms wear ***taxpayer*** hats. The taxes they pay to government are allocated as needed to fund the bonds and grants that provide the capital to build the facility, to cover its annual operating costs, and to support all other public services and facilities.
- In the public-private partnership approach, individuals and business firms wear ***customer*** hats. They are charged commercially viable prices for access to the facility. The business-type revenue this produces flows through the partnership and is dedicated exclusively to covering the facility's operating costs and supporting its capital structure.
- In the all-public approach, the operating and capital needs of all government undertakings are ***co-mingled*** and funded from a ***single*** revenue source called tax collections.

- In the public-private partnership approach, ***each infrastructure facility stands alone.*** It generates its own revenue by charging fees to its customers, and is responsible for paying its own costs. There is no co-mingling of either revenues or costs with various other facilities.

The all-public approach embodies classical public-finance theory. This can work quite effectively in certain kinds of societies. In Hong Kong, for example, the government imposes a high public-savings rate on residents and business firms through its tax policies, and then uses its resulting "operating budget surpluses" to fund the construction of infrastructure facilities on a largely "pay-as-you-build" basis. Such expanded infrastructure capacity is a key element of the government's strategy for guiding metropolitan development in directions that promise greater economic prosperity in the future. In Hong Kong, this has contributed significantly to a level of per capita gross domestic product that is one of the highest in the world.

But classical public-finance theory seems to have broken down in the United States. Americans find it increasingly difficult to perceive a valid link between the value of the benefits they receive from government undertakings and the size of their tax bills. Too often, they suspect that the "other guy" is receiving too large a share of the benefits while they're being saddled with too large a share of the costs.

These suspicions have contributed to a growing national unwillingness to finance new infrastructure projects with taxes. But at the same time, they have also increased the perceived attractiveness of user charges like tolls, which are regarded as being "more fair" because only those who actually use a highway for other public facility are required to pay for it.

Conclusion

Now that we have the technology needed to apply what is essentially movie-ticket pricing to our roadway systems, we can significantly improve the way we operate and fund them.

- We can make these systems fully self-supporting by charging fair prices to those who actually use them.
- We can operate them more rationally by varying these prices to reflect demand. On each individual roadway link. At different times of day. By each class of motor vehicle.

- We can expand roadway capacity to meet tomorrow's travel needs by creating formal partnerships between business and government to build and operate new roadway links.
- These public-private partnerships can raise the necessary construction funds from a much wider spectrum of capital sources on the strength of their revenue streams generated by motorists and other customers.

We can call this a revolution if we wish. Or can simply see it as another example of America's natural talent for developing new solutions to new needs. Either way, the time has come to take it seriously and start putting it to work.

THE POWER OF TECHNOLOGY

Creating Customers

In his landmark 1973 book, *Management: Tasks, Responsibilities, Practices*, social scientist and management guru Peter F. Drucker insisted that the only valid purpose of a business enterprise is to create customers.

The customer is the foundation of any business and keeps it in existence. Only the customer can determine what products an enterprise should produce and how they should be sold. So the true definition of entrepreneurship (in its most exalted Austrian School sense) is the process of creating customers.

But what about profits? Haven't we always been led to believe that the purpose of a business firm is to make money?

Drucker's response to this is both simple and profound. Profit is simply one of the costs that an enterprise has to cover, like employee salaries and payments to suppliers. So defining the purpose of a business enterprise in terms of maximizing profits is ultimately self-defeating, just like defining its purpose in terms of minimizing employee salaries or supplier payments. After all, both can be reduced to zero by closing down the enterprise. What really counts is creating customers.

One of the things we must do to create customers is to find ways of maximizing the trade-off in each customer's mind between the value that he or she perceives from buying our products and the cost to him or her of obtaining these products. An example of how this works is illustrative.

Interestingly enough, this example doesn't come from the private sector. It does not involve Microsoft, General Electric or any other highly successful corporation. The following example, believe it or not,

---

Adapted from Joseph M. Giglio, "Creating Customers," Public Works Management & Policy *7, no. 4 (April 2003): 231–234.*

comes from the much-maligned public sector, which many people believe is never a good example of anything except klutzy incompetence. The example is the New York City transit system.

By any measure, New York City's subway and bus system is far and away the largest urban transportation enterprise in the United States:

- Its current total of 2.1 billion annual paid transit trips exceeds the combined total of the next five largest urban transportation systems.
- Its 260 subway and bus routes cover 1,900 miles.
- Its 40,000 employees operate and maintain 10,200 subway cars and buses.
- It is a far larger factor in providing daily mobility for its city's eight million residents and two million daily commuters from the surrounding suburbs than any other American transit system.

During the past hundred years, it has had a much greater influence in shaping the New York City we know today than other American transit systems have had on their respective cities, from the immense clusters of office towers that define the very shape of the earth in Manhattan's Central Business District to the unimaginably vast sweep of apartment buildings and row houses that fill hundreds of residential neighborhoods in Northern Manhattan, Brooklyn, Queens and the Bronx. The New York subway system's impact has been immense.

And it operates all day, every day, without pause throughout the year, as inexorably as the world turns, in the city that never sleeps.

All of this provides a fresh sense of reality to the oft-used term "monumental." But equally monumental are its efforts during the past five years to create what amounts to some half-billion new customers.

It has done this by using modern technology to change the way its customers think about buying and paying for trips on subways and buses. This has resulted in a 27 percent increase in subway ridership and a 50 percent increase in bus ridership, for an aggregate one-third increase in total annual ridership.

Until the mid-1990s, New York City's transit system followed the traditional practice of charging its customers for each trip they made. That is, a customer would buy a group of tokens at a subway station change booth. They would then either drop a token into a station turnstile to make a trip by subway or into bus fare box to make a trip by bus. Therefore, they bought and paid for each trip as a separate and distinct entity.

However, this is not how people think about trip making. Survey research shows that they regard trip making in origin/destination terms. For a person who lives in the Brooklyn neighborhood of Marine Park and works in Rockefeller Center, his or her commuting trip origin is the row house in Marine Park and the trip destination is Rockefeller Center. Therefore, he or she thinks of the commuting trip as a continuum that connects the house to Rockefeller Center.

But when the commuter actually makes this trip, it turns out to involve three separate legs. First, there's the short walk to the bus stop on Avenue U (which costs nothing). And then there's the bus trip to the Avenue U subway station on the Brighton Line (which costs one token). And finally, there's the subway trip to the Rockefeller Center station at 47th Street and Sixth Avenue in Manhattan (which costs a second token).

Separating this commuting trip continuum into three legs (each of which involves a different travel mode) forces the traveler to think in terms of three distinct trips. Because two of these trips must be paid for with one token each, the traveler has to be aware of how many tokens he or she has.

In a broader sense, therefore, a New Yorker's use of the transit system for any trip continuum is influenced by an awareness of how many separate transit mode trips may be involved. Because each of these trips must be paid for separately, using the transit system for a major portion of the trip continuum can involve multiple payments. So the traveler becomes conditioned to think in terms of a series of trade-offs:

- Between walking to the final destination versus taking a bus or subway train.
- Between walking to a subway station versus taking a bus to the station.
- Between taking a less direct and more time-consuming route to the final destination to use only one transit mode (and therefore pay only one token) versus saving time by taking a more direct route involving two or more transit modes (which means paying two or more tokens).
- And, more significant in terms of overall transit ridership, between making the trip by transit versus your own car, or even between making the trip at all versus not making it.

These trade-offs are influenced by the traveler's perceptions of cost, because the number of transit trips he or she makes determines the number of tokens that must be spent. And these perceptions influence the ultimate trade-off, which is between value and cost. So the traveler instinctively weighs the value he or she receives from using transit versus the cost of using one or more transit modes for the bulk of his or her trip continuum.

In the mid-1990s, New York City's transit system began installing a totally different system to charge its customers for subway and bus trips. This system, known as MetroCard, involved selling transit customers prepaid electronic fare cards that could be used in any subway station turnstile or bus fare box instead of old-fashioned tokens. Its original attraction was its ability to reduce the operating costs involved in selling tokens and increase the size of the transit-fare float (which could be loaned out overnight to earn interest income).

Because MetroCard's technology was fairly complex and quite different from anything the transit system had used previously, it took several years for the system to become fully operational. This created opportunities to do other things with MetroCard, which come under the heading of Smart Marketing. As a result, MetroCard came to include some interesting features:

- Customers can buy MetroCards from automatic machines at each of the city's 468 subway stations. They can charge the price to one of their credit cards, so they don't actually have to lay out any cash until the payment due date for next month's credit card bill.
- Customers can select the number of transit rides they purchase. For every ten rides purchased, they receive one extra "free" ride.
- MetroCards give customers free transfers between buses serving different routes. This means that savvy New Yorkers traveling in areas with numerous bus routes can often make a round-trip for the price of a one-way trip by using different routes going and returning.
- MetroCards also give customers free transfers between bus routes and subway lines. This reduces the cost of using multiple transit modes to complete a door-to-door trip continuum.

These features enable MetroCard to embody a classic principle practiced by smart retail business firms everywhere: "If you can divorce the act of buying from the act of paying, you will sell more to each customer."

This principle works in retail stores that accept credit cards because customers' purchases are no longer limited by how much cash they have on them. They may go to the store to buy a single item. But once in the store, they see several other items that attract their attention. So they buy them too, and charge everything to their credit card. The limit on how many impulse purchases they make is often determined by how large a bag of items they are able to carry home.

It turns out that this principle works just as effectively for transit systems. With a MetroCard, New Yorkers can decide to buy trips by bus or subway based on impulse, how late they are running, or how tired they are, rather than on how many old-fashioned single-trip tokens they have in their purse.

Full operation of the MetroCard system has brought about a wholesale change among New Yorkers in their perceptions of the trade-off between value and cost in using transit for trip making. Awareness of the cost per trip has been greatly diminished by not having to part with "cash" (U.S. bills and coins or transit tokens) to make each transit trip. This has heightened awareness of the value of being chauffeured around town by bus drivers and subway motormen (and not having to worry about parking or being able to flag down a taxi in the rain).

All of this is reflected in the one-third increase in annual paid subway and bus paid trips during the past five years, even though New York City's population increased only slightly during this period.

This obviously indicates big-time success in following Peter Drucker's dictum of creating new customers—in this case, by exploiting a technology that was originally seen mainly as a way to reduce operating costs.

THE POWER OF TECHNOLOGY

Beyond the Token: CharlieCard, Service Improvements and Customer Perceptions

In June of 2005, a Bostonian returning from a trip out of town would have encountered something new at the Logan Airport subway station. In the entryway was a line of shiny new machines dispensing something called a CharlieCard. If this traveler had spent much time in cities like New York or Washington, D.C., she would have breathed a sigh of relief that Boston's transit system had finally arrived in the 21st century. Gone were the days of tokens—to be stocked up on every week, fumbled with at the turnstile, and, too often, lost along the way—replaced with a simple, refillable debit card. But this was just the first stage of the CharlieCard roll out; it would work on one subway line and one bus line. For now, the traveler would have to save that new card for the next trip to the airport.

This article discusses three questions that are of interest to the Massachusetts Bay Transportation Authority (MBTA) and the people of the Boston metropolitan region as the MBTA enters a new era:

- How can the MBTA increase the use of its new CharlieCard?
- How can the MBTA increase ridership on the transit system?
- How do passengers view transit-system performance?

The answers to these questions are not simple, but they are tractable—in the sense that they are drawn from the larger world of human experience beyond MBTA headquarters and the Boston region as a whole. Although this article is far too brief to explore them fully, its discussion can point the way to the kind of thinking and analysis needed to produce answers that are true solutions.

Adapted from Joseph M. Giglio, "Strategic Notes: CharlieCard, Service Improvements, Customer Perceptions" (report prepared for the Massachusetts Bay Area Transit Authority, August 8, 2005).

How Can the MBTA Increase the Use of its New CharlieCard?

The CharlieCard involves new technology, and new technology has a tendency to make people nervous. Using it means that they must change their established habits, which they are generally reluctant to do. So they seek comforting excuses to ignore new technology for as long as possible, hoping that it will eventually go away, which is what happens more often than not. Overcoming this barrier is a major challenge to widespread acceptance of the CharlieCard among Boston transit riders.

Fortunately, other large transit operators in the United States that have adopted various forms of automatic fare collection have met this challenge successfully. They include New York's Metropolitan Transportation Authority, Washington's Washington Metropolitan Area Transportation Authority, Chicago's Chicago Transit Authority, and San Francisco's Metropolitan Transportation Commission. The MBTA can make things easier for itself by analyzing the experiences of these other transit operators in detail to determine which aspects are most applicable to the Boston environment.

This analysis requires an understanding of two issues. One is the innovative nature of new technology, which is central to the benefits it offers. The second is the process through which products embodying new technology diffuse (or fail to diffuse) throughout their target populations. Of these, the second is the most important.

To truly succeed, new products embodying innovative technology must come to be accepted by the majority of their target customers. If they gain only minimal acceptance, they become niche products or simply fall by the wayside.

For a product like the CharlieCard, something close to universal market acceptance is required if its broad vision is to be a success. But the majority market will only consider adopting new technology after innovative users have tried it and given it their thumbs up.

The order in which the whole spectrum of target customers adopt new technology products is termed the **DIFFUSION PROCESS**. The idea is to move as quickly as possible to capture the **INNOVATORS**, then pursue the **EARLY ADOPTERS**, the **EARLY MAJORITY**, the **LATE MAJORITY** and finally the **TRADITIONALISTS** (who may never become fully comfortable with the new technology). The goal is to make the product seem ubiquitous.

The length of time for new technology products to become accepted by the Late Majority may not be as short as we would like. The issue for the CharlieCard is whether its vision can be broken into easily digestible pieces. This raises the question of which portions of the system can be implemented before the whole system is up and running on all buses and trains. As the MBTA moves towards establishing a truly ubiquitous system, which customers can it hope to capture along the way? Customers want to see clear value and little risk. They want something of value to them personally.

We can appreciate how this works by remembering the way personal computers moved from home-assembly kits sold by small mail-order companies to the Heathkit crowd of venturesome hobbyists in the early 1970s to today's standard home and office appliance that can be bought virtually everywhere along with an immense variety of user-friendly applications software.

The key to this widespread diffusion of a truly "super-technology" product involved weaning large segments of the general public away from their sci-fi perception of computers as huge, mysteriously complex machines guarded by an imperious priesthood of techno-geeks who spoke a language only vaguely similar to English. Instead, the public had to be coaxed into perceiving personal computers as useful tools whose technological details were unimportant to the average Joe, who only cared about how a computer could make his life simpler and more productive.

Like all successful new technology products, five factors accounted for the rapid diffusion of personal computers into widely used appliances taken for granted in today's homes and offices.

- ***Relative Advantage*** (the extent to which the new-technology product is better than the old product it can replace). For example, personal computers greatly increased the productivity of office secretaries, who could now quickly and easily revise the boss's letters and memos right on the screen instead of having to retype entire pages. And the automatic spell checkers built into word-processing software ended the time-consuming process of having to thumb through dictionaries looking for the correct spelling of words. These productivity gains weren't lost on their bosses.
- ***Compatibility*** (the degree to which the new product is consistent with existing values and experiences). Personal-computer keyboards are

virtually identical to standard typewriter keyboards. So anyone who can type has already passed first base in learning how use a computer.

- ***Complexity*** (how difficult it is to understand and use the new product). Mail-order buyers of the early home-assembly personal computer kits had to know how (or learn how) to write programs in higher languages like Basic or Fortran if their new computers were to be something more than interesting living room displays, and this was scarcely everyone's cup of tea. But today's personal computer buyer is actually purchasing access to easy-to-use software for word processing, building spreadsheets, creating simple or complicated pictures, accessing the Internet and doing other things that interest him. The computer itself—including the remarkable technology it contains—is incidental.
- ***Try-ability*** (the degree to which the new product can be experimented with on a limited, low-cost basis). During the early 1980s, IBM's breakthrough PC with its applications software for word processing and spreadsheets quickly penetrated the office market. This made it relatively easy for young managers and other office workers to try them out at odd moments, at no cost, and under low-pressure conditions. Positive experiences led many of these people to request PCs for their work and buy them for use at home.
- ***Observability*** (how easy it is for non-users of a new product to see it being used by others and become aware of its benefits). As managers saw how quickly and easily their secretaries could type letters and memos on personal computers, they began thinking about others ways PCs could be used to improve staff productivity—and even their own.

By understanding how these factors apply to the CharlieCard and learning how to manage them effectively, the MBTA can tailor its marketing efforts to exploit the technology's value to riders. It can also gain a better sense of how long the CharlieCard's diffusion process may take and identify the points at which it can be accelerated.

In a perfect world, the MBTA could wave a magic wand and immediately have the entire transit system fully wired so that the CharlieCard could be used on every bus and every train. But this is clearly impossible. Installing CharlieCard technology throughout the system is going to be an incremental process that could take some time. In the meantime, the present fare-payment systems will have to remain operational.

This confronts the average rider with an interesting choice. If he wants to use the CharlieCard, he must be prepared to also use the other fare-payment systems for some period of time. Given this reality, many riders will understandably elect to continue using the other fare-payment systems with which they are already familiar and not bother with the CharlieCard. And as the fiasco of introducing the metric system in the United States has demonstrated, the existence of parallel systems can often discourage the acceptance of more innovative systems.

Under these circumstances, the MBTA's best course for increasing the use of the CharlieCard is to follow three complementary strategies:

First Strategy: Continue a full-court press with the assistance of Schien Bachman to complete installation of the CharlieCard on every bus and train as quickly as possible—with completion time measured in months rather than years.

Second Strategy: From the beginning, provide meaningful incentives for riders to use the CharlieCard wherever it is accepted. Doing so accomplishes two things. First, the right kind of incentives can offset the disincentive of having to be ready to use several other fare-payment systems if one wants to use the CharlieCard. Second, providing incentives to use the CharlieCard imposes an implicit disadvantage on the use of other systems. While it might be simpler in the abstract to pass laws that ban the use of the other systems wherever the CharlieCard is accepted, this is unlikely to happen in the real world. CharlieCard incentives (coupled with artful administrative barriers to using the other systems) are probably the most practical way to help speed the day when the CharlieCard is the only game in town.

During the installation period for its Metro Card system, New York's Metropolitan Transportation Authority offered Metro Card users (but not riders who paid with tokens or cash) free transfers between buses and subway trains and from one bus line to another. This free transfer policy is regarded as an important factor in the relatively rapid increase in bus and subway riders using the Metro Card and has remained in effect ever since. It was also accompanied by a strong increase in total ridership as transit service quality experienced significant improvements.

Obviously, this kind of incentive has its trade-offs. Increases in ridership from free transfers can also result in operating deficit increases. Whether these trade-offs are positive may depend on whether the MBTA can convince the powers-that-be that temporary increases in operating deficits should be regarded as worthwhile investments in a more transit-friendly future.

Once CharlieCards become as easy to use as the other fare-payment systems (but far more convenient) on all MBTA buses and trains, it is a relatively simple matter to phase them out if the MBTA wishes. This won't matter to most riders, but it can save the MBTA the high cost of processing cash receipts. This was the experience of New York's MTA with its Metro Card, though the impact on ridership from the elimination of tokens was not material.

Third Strategy: Offer CharlieCards that are automatically refillable so each rider can choose the option of never again having to worry about whether his card has enough stored credit to pay for a ride. The way to do this is to issue permanent cards at no cost to all who want them. Each card would be linked to a computerized account in the cardholder's name. The cardholder can then authorize this account to automatically add a specified sum to the card balance whenever it falls to a certain minimum level, with the sum added being charged to one of his major credit cards.

The end result is that the cardholder always has enough credit to ride the system whenever he wishes and as often as he wishes. From the standpoint of pure convenience, this is the next best thing to establishing a no-fare transit system. At the same time, automatically refillable cards tend to make future fare increases more of an abstraction to cardholders.

(In New York, the MTA only makes automatically refillable Metro Cards available to senior citizens who qualify for special half fares because of their age. The reasons for this restriction are not clear, but they have the effect of removing another incentive to use the Metro Card to ride more often.)

HOW CAN THE MBTA INCREASE RIDERSHIP ON ITS TRANSIT SYSTEM?

The standard view in the transit industry and among transit advocates seems to focus on two strategies for increasing ridership.

The first is to expand the system's market base by opening new bus and train lines that provide service to areas of the metropolitan region that have not previously been served. In theory, this increases the total number of riders who can use the transit system (though only a minority of them probably will).

The second strategy is to upgrade the quality of service on existing lines across the board so that riders and potential riders perceive the system as being competitive with what private automobiles are presumed to offer in travel safety, reliability and convenience.

These two strategies are competitive. The resources used for expanding the system are obviously not available for upgrading service quality on existing lines, and vice versa. While in a few exceptional cases a metropolitan region's powers-that-be may dedicate to one strategy certain resources that are not available to the other, the choice between the two is usually wide open. This means transit-system managers must make policy decisions about which strategy to follow and convince the region's powers-that-be that their decision is the correct one. Whichever way they go, such decisions will have implications for the amount of management talent required as well as for the capital and operating dollars required.

Opening new lines is often a popular choice because it allows a region's power-that-be to provide new benefits to certain segments of the region's population that are in a position to command attention. Transit managers who meet these demands often find doing so makes their lives easier in other ways.

But opening new lines carries with it an inevitable rise in the system's total operating costs. And the historical record among transit systems in the United States indicates that the increase in systemwide operating costs arising from expansion projects is usually not covered by increases in systemwide revenues. This can lead to problems, especially when narrow by-the-book accounting practices that ignore economic externalities are involved.

In certain cases, communities hoping to receive transit service for the first time must be asked to provide higher levels of tax support for operating costs than is the norm for communities that already have service. This may cause the communities affected to re-evaluate how important transit service really is to them. Or it may cause them to voice elaborate and noisy rationales about why it is unjust to "penalize" them now simply because they didn't receive transit service years ago when other communities did.

In the rare cases where the only clear alternative to new transit service is significant new road construction, artful transfers of public funds from roadways to transit may be possible—possibly under the justification that "it's cheaper and less environmentally intrusive for the state government to use funds it would otherwise spend on new roads to support new transit lines." But gaining approval for such transfers is a complex task and often turns on issues other than transportation merit.

The Hong Kong Model

Perhaps the ideal environment for providing superior transit service is to be found in the semi-autonomous Chinese metropolitan region of Hong Kong, whose nearly seven million residents enjoy a level of per capita gross domestic product that is exceeded only by the United States and Norway. During the past decade, Hong Kong has spent more than US $20 billion to expand its subway and commuter rail systems and has begun work on another US $13 billion program to expand these systems further by 2010. This 20-year total of US $33 billion far exceeds the combined total for all new U.S. subway and commuter rail lines built during this period.

At the same time, Hong Kong invests large annual sums in its existing subway and commuter rail lines to enhance the quality of service to passengers, improve the productivity of workers and help keep operating costs under control. This enables these systems to maintain their reputations for providing transit service that is generally regarded as the best in the world (complete with subway stations that are fully air conditioned and feature up-and-down escalators from the street to the train platforms).

Hong Kong's subway and commuter rail systems are operated by two commercial corporations owned by its metropolitan government. Their expansion projects are largely funded by new equity investments from the government plus judicious issues of debt, while on-going capital improvement programs are supported by investments of accumulated profits.

The mention of "accumulated profits" may raise questioning eyebrows among many Americans. But these two companies actually make money from the fare box by carrying passengers, so there are never any issues about taxpayer subsidies. This is largely because public transportation in Hong Kong accounts for an astonishing 88 percent of all daily non-pedestrian trips, thanks to comprehensive land-use policies that favor travel by transit.

Hong Kong is able to support this kind of ambitious capital investment in transit service because its government is far richer than any government in the United States. There are two reasons why:

The government controls all the land in Hong Kong. Therefore, all significant real-estate development projects originate within the government's top-down central planning process for regional development rather than in the offices of private firms. The government leases land parcels to private developers on a long-term basis in accordance with its planning process (which, among other things, favors increased use of rail transit for local travel). The considerable income from these leases is largely dedicated to investments in new transportation and other infrastructure that will make Hong Kong an increasingly attractive place to do business.

The government has a monopoly on all tax collections in Hong Kong. If you work or own a business in Hong Kong, you pay taxes only to the metropolitan government—not to any state or national government. This means that your total tax burden is lower than in other cities of the developed world. But it also means that the per capita tax collections of Hong Kong's government can be considerably higher than for governments in U.S. metropolitan regions. During the 1990s, for example, this meant that the operating revenues of Hong Kong's government regularly exceeded its operating expenses by a good 20 percent. In effect, the government was imposing a relatively high public-savings rate on Hong Kong's economy in order to generate funds to invest in transportation infrastructure and other programs to enhance Hong Kong's prosperity.

The ability of Hong Kong's government to generate lavish amounts of funds to invest in transportation facilities and other new infrastructure to stimulate future economic growth may be a source of envy for the rest of us. But that is not the point. What we should realize is that Hong Kong represents an idealized textbook model for how metropolitan regions like Boston should be managed in order to make maximum contribution to American prosperity. However, this is not a model that can be implemented as it stands in the United States. The prevailing view here seems to be that metropolitan regions should function as cash cows for federal and state programs that try to maintain the economically dysfunctional aspects of rural life.

The Reality in Boston

All of this may raise interesting issues for classroom debates at the Kennedy School. But they have little to do with the bread-and-butter issues that have to be decided in a metropolitan region that lacks Hong Kong's financial autonomy and enormous government wealth. And among those issues in Boston are policy questions about the most appropriate direction for the MBTA to pursue in allocating all-too-scarce capital dollars among competing transit needs.

Perhaps the most appropriate metaphor in this case involves the homeowner who is trying to decide between adding an extra room to his house or replacing an antiquated heating system that broke down four times last winter—always on especially cold days and always requiring expensive repairs. To most people, the homeowner's choice is a no-brainer. He must restore the heating system first (along with the leaky roof, if that's also a problem), and postpone adding the extra room until some future time. Otherwise he is courting disaster.

If MBTA managers wish to turn around recent ridership declines and avoid paralyzing debates about the future fare increases that are inevitable, they must provide transit service of acceptable quality to Boston's bus and train riders. Some estimates place the capital cost of restoring the transit system to a state of good repair at nearly $700 million per year for the next seven years, plus about $480 million per year thereafter to keep it that way. This amounts to an ongoing 20-year capital program totaling more than $11 billion if no new transit lines are added. Apart from improving service quality for passengers, such a program will also help kept operating costs under control by eliminating the need for costly emergency repairs and by raising worker productivity.

It is estimated that the MBTA is currently spending slightly more than $400 million per year on the capital needs of the existing transit system. Over 20 years, this amounts to about $8 billion, or less than three-quarters of what is needed to restore the system and maintain it in a condition capable of providing decent transit service. This implies that a continuation of current spending levels inevitably means deteriorating service quality, declining ridership, more frequent emergency repairs that raise operating costs and a generally dim future for Boston's economy.

This is similar to the challenge that New York's Metropolitan Transportation Authority had to confront in the early 1980s, and there is much to learn from how it met this challenge.

In summary, service quality had been deteriorating for many years on the MTA's buses and trains due to deferred maintenance and a seriously under-funded capital replacement program. Inevitably, this meant ridership declines. But no one in the political community paid much attention to the ongoing clamor this raised among transit advocates.

It took a new MTA chairman drawn from the local business community to change this. By means of clever salesmanship, he convinced the leaders of the business community that transit in New York faced a service crisis that threatened their corporate bottom lines. With the active support of these leaders, he negotiated a multi-year capital funding commitment from the New York state legislature large enough to begin the process of restoring the MTA's bus and train systems to a state of good repair. Since the early 1980s, the MTA has spent more than $40 billion on this task and its service quality is now regarded as among the best in the nation.

But more than money was required. Like all older transit systems, the MTA had inherited certain traditional work rules that prevented its managers from fully exploiting the benefits of new capital improvements. These had to be changed if riders were to enjoy better service quality.

An obvious example involved transit workers who maintained the air-conditioning systems on buses and trains. These workers were civil servants who were entitled, like all New York civil servants, to apply for vacation time during any portion of the year. Since most of these workers had families, they naturally chose to take their vacations during the summer, when air-conditioner maintainers were most needed. This left the MTA chronically short of enough maintainers to keep bus and train air-conditioning systems working properly during the hot summer months. So a large proportion of the buses and trains had non-working air conditioning.

The MTA resolved this problem by obtaining authorization from the State legislature (again with the support of the business community) to create a new job category for air-conditioner service workers. New workers hired for these jobs would enjoy the same pay scales and benefits as existing workers, but they would not be civil servants. And they could not take vacations during the summer. The MTA then proceeded to hire new air-conditioner maintainers only in this new job category and ceased any hiring for the old civil-service category. The result was that the MTA soon had enough air-conditioner maintainers working during the summer months to assure that all buses and trains had working air conditioners.

Eventually, the two job categories were merged into a single civil service category. But the ban on summer vacations remained, and New Yorkers have come to take for granted comfortably cool buses and trains during the summer as simply one more fact of life. A similar approach was followed in dealing with other work rules found to limit improvements in service quality.

Once an urban transit system falls into the slough of despondence that New York's MTA had to confront in the early 1980s, glitzy new fare-payment systems may seem beside the point. But they are actually part of bringing a transit system into the 21st century. An aggressive capital restoration program can do more than simply return to us the service quality that we enjoyed a generation ago.

An important component of good service quality concerns how ***user-friendly*** the system is. Once passengers are able to take decent-quality bus and train rides for granted again, the CharlieCard can be one of the things that makes people want to use Boston transit more often. The reason why involves a profound truth about mass-market retail success that department stores like Filene's first stumbled onto many years ago:

If you can separate in the customer's mind the act of **BUYING** ***from the act of*** **PAYING*****, the customer is likely to buy more each time he's in the store.***

Once Filene's and other individual department stores realized this profound truth, they developed the individual charge card systems that enabled their customers to buy today but not pay until the end of the month. The increases in retail sales this brought about eventually paved the way for the rise of the huge general credit-card industry that greatly simplified retail shopping for millions.

The psychology underlying this concept is not difficult to understand. A customer walks into Filene's to buy a particular product. But once inside the store, he finds himself surrounded by vast numbers of other attractive and good-quality products he may not have considered buying before. If he's not constrained by the amount of cash he happens to have in his pocket, he may very well buy one or more of these other products in addition to the product he originally came to the store to purchase.

Filene's charge card (and credit cards generally) remove the constraint on buying today by removing the requirement of paying today. In fact, the only remaining constraint on how much a customer buys today may be how much he can carry home. And at the end of the month when he receives his charge-card statement, he writes a check for the total amount without spending much time looking at the individual items that make up this total.

The MBTA is in the same kind of mass-market retail business as Filene's, even though it sells transit rides rather than toasters or shoes. So the same formula for success in the retail business holds true. If the MBTA can separate in the customer's mind the act of buying transit rides from the act of paying for them, it is likely to sell more transit rides. The CharlieCard can make this separation process feasible once it is accepted on all buses and trains in a system whose service quality is good enough to be taken for granted.

Think of a commuter who lives in Cambridge heading to a nine o'clock business meeting in downtown Boston on a January morning five years from now. As he trudges through the snow along Massachusetts Avenue towards the Harvard Square subway station, the bitter cold wind turns his face into a frigid mask of agony (because it will still be a fact of life five years from now that Boston's January winds always blow against the direction you're walking).

Then the commuter glances over his shoulder and sees a new bus proceeding behind him down Massachusetts Avenue towards Harvard Square. He knows that the bus is going to stop at the next corner. He knows that these buses have especially effective heaters. If he has an automatically refillable CharlieCard in his pocket, isn't it likely he's going to board that bus as soon as it stops? Why not? It isn't as if he has to pay for the bus trip with cash in his pocket. Therefore, its cost is simply an abstraction. What isn't an abstraction is relief in a warm bus from that terrible January wind.

This may seem like an extreme example, and it probably is (though Bostonians who know about the Arctic severities of January may disagree). But what isn't extreme is the number of additional times per day that CharlieCard holders are likely to use reliable MBTA buses and trains for discretionary trips because they feel that they've "already paid for them." So why not make life easier, as credit cards have taught us to do?

The various versions of the CharlieCard can provide a long step in the direction of separating the act of taking a transit trip from the act of paying for it. However, the kind of automatically refillable CharlieCard described in the previous section can make this separation complete. You don't even have to write a check at the end of the month when your statement arrives, as you must with regular credit cards. So taking MBTA buses or trains becomes as natural and unthinking as walking. If a reliable bus or train is available, just take it and make life easier.

How Do Passengers View Transit System Performance?

Standard approaches to conducting passenger surveys don't always provide meaningful answers to this question. If you ask a passenger point blank what he thinks of the system's performance, chances are he's going to answer "TERRIBLE." Just as if you ask him what he thinks of Boston's weather, he's also going to answer "TERRIBLE."

The reason in both cases is simple. Most of the time, passengers pay no attention to their transit trips (or to the weather). This is because the average trip is made without any problems, just as most of the time Boston's weather isn't uncomfortable. So there's nothing about the trip to grab the passenger's attention.

The only time a passenger pays any attention to his transit trip is on those occasions when something disagreeable happens. The bus or train is noticeably late in arriving. Or its air conditioning or heating system isn't working. Or there's a major delay during the trip that causes the passenger to arrive at his destination noticeably late. These events stick in the passenger's mind and cause him to form the impression that transit service is "terrible"—just the way a rare heavy snow storm in January or heat wave in July cause him to form the impression that Boston's weather is "terrible."

This may be why passenger perceptions are not fully captured by the MBTA's existing performance statistics. If the reporting system indicates that 90 percent of a day's total end-to-end runs were completed "on time," this figure simply measures the peak of the bell-shaped curve along which the timing ratios for all trips are distributed. It tells us nothing about how widely these ratios are distributed. Those ratios that fall to the left of the curve's peak (i.e. "late trips") are what matter most in determining passenger perceptions because they are the ones, like the rare January blizzard, that passengers most remember. So the negative perceptions created by a single late trip can offset the positive perceptions created by a great many on-time trips.

CONCLUSION

The MBTA faces two interesting challenges as it considers the three questions discussed here:

- Restoring Boston's transit system to a state of good repair before it is too late, so that passengers once again enjoy the kind of decent-quality public transportation they can take for granted.
- Bringing a fully restored system into the 21st century by using the CharlieCard to make fare payment the kind of user-friendly and transparent process that no longer causes riders to think twice about traveling by bus and train.

In deciding how to meet these challenges, the MBTA can stand on the shoulders of others in the transit industry who have had to manage extensive system-restoration programs and make automatic fare-collection technology work effectively on their systems in both technological and human terms.

Boston may not be Hong Kong in terms of its available resources for better transit, but it remains one of the nation's ten largest metropolitan regions in terms of gross domestic product (which matters more than raw population). And its transit system should once again be one of the nation's best.

SELF-SUPPORTING ROADWAYS

New Horizons in Roadway Financing for Major Metro Regions in the United States

In retrospect—and by conveniently ignoring some important physical and institutional constraints—it is easy to conclude that the United States has made a series of wrong-headed choices about how to finance its all-important metropolitan roadway systems. The results of these mistakes are all around us. They take the form of:

- Too little roadway capacity where it is most needed, as evidenced by severe traffic congestion on many critical roadway links in our most important metropolitan regions during increasingly long portions of the day.
- A chronic inability to build new roadway capacity to keep up with demand, even to the point where blindly chanting that "we can't build our way out of congestion" too often replaces serious discussion of how to overcome obvious capacity shortfalls.
- Ingenious proposals for repackaging available capital funds in artful ways, which may enrich investment bankers in the near term but do little to increase the flow of funds over the long term.
- "Saving money" in government operating budgets by reducing maintenance on roadway systems in the near term, causing them to wear out faster and lose some of their existing capacity in the long term.
- Failure to actively manage our roadways in sensible ways so that we maximize their functional capacity and provide the most service to the most travelers.

Adapted from Joseph M. Giglio, "New Transportation Initiatives and Demands on Financing" (speech, Transportation Research Board Third National Conference on Transportation Finance, Chicago, Illinois, October 27–30, 2002).

To move beyond these mistakes, we must first separate myth from reality when it comes to roadway financing. More importantly, we must recognize the potential of recent technological breakthroughs to form the basis of a radically new and more effective market-oriented roadway financing structure.

TRANSPORTATION AND ECONOMICS: MYTHS AND REALITIES

Economic activity generates travel demand. Workers must commute between their jobs and homes, factories must ship their output to customers and receive shipments of raw materials and supplies, salesmen must call on customers, service and repair personnel must travel to the sites where they perform their work, and so on. Because the relationship between economic activity and travel demand is positive and roughly linear, increases in economic activity are likely to result in higher travel demand. And in U.S. metropolitan regions, most of this travel demand must be accommodated by the roadway system. Therefore, increases in daily traffic flow on these systems are a measure of a growing economy.

All this may seem too obvious to dwell on. Yet there is an astonishing lack of understanding of its significance among certain environmentalists and other anti-road types who have come to strongly influence decisions about expanding roadway capacity in many U.S. metropolitan regions. In fact, leading figures in these circles have become so obsessed by a phenomenon they call INDUCED TRAVEL DEMAND that they have brought new road building to a halt in many regions.

Induced Travel Demand is nothing more than the not-surprising ability of new roadway links to generate trips that were not previously made. What road-building opponents fail to realize is that very few of these new trips come under the heading of pure, wasteful joy riding. Virtually all of them have some economic rationale, so that their occurrence means an increase in economic activity.

So the ability of new roadway links to generate new trips can legitimately be called INDUCED ECONOMIC ACTIVITY. This is something that presumably should be welcomed by anyone who believes that rising incomes and wealth are positive social outcomes (though this view is obviously not shared by those who have adopted a Marie Antoinette vision of society).

Therefore, let us keep in mind a ruling principle that may seem too obvious to belabor (but clearly isn't):

A society's transportation systems, including its roadways, exist to support its economy. Not vice versa.

The Pricing Problem

Economic activity generates demand for more than just travel. It also generates demand for a whole host of other things, including labor, commercial space, raw materials and supplies, electricity and other utilities, and a wide range of services. The significant difference is that these other things can usually be supplied through some kind of market system.

While market systems may be something less than the roads to Nirvana that their libertarian advocates like to claim, they do offer some compelling benefits. Not the least of these is the potential for willing buyers and sellers to exchange various commodities in a market environment at prices that are largely determined by the ever-changing relationship between supply and demand. The objective nature of these prices provides valuable economic signals to market participants—helping buyers determine what quantity of a commodity it makes sense for them to risk obtaining, and helping sellers determine what quantity of a commodity it makes sense for them to risk providing.

The daily ebb and flow of prices in such markets thus structures an objectively rational complex of systems for distributing commodities throughout society, moving them more or less smoothly from the hands of producers to the hands of consumers. In general, this works so well and with such astonishing sophistication that one can easily imagine it being the invention of some highly gifted economists who munch on differential equations between meals.

In fact, it is an entirely natural phenomenon. It arises from nothing less than the twin human instincts to trade with each other and to invent monetary systems that facilitate trading—instincts that seem to be hard-wired into the DNA of the human species at least as solidly as the instinct to reproduce.

However, the unique characteristics of certain commodities make them unsuitable for distribution through the kind of marketplaces with which we are familiar. Metropolitan roadway systems are an obvious example. Until recently, it has rarely been practical to subdivide the services they provide ("lane miles for travel purposes") into conveniently small units, to price access to these units in some objective fashion, and to charge buyers for their use. The special conditions present in bridges and tunnels on which tolls are charged illustrate the logistical challenges of trying to price access to metropolitan roadways the way we price access to such commodities as seats in movie houses.

Fortunately, technologies now exist that enable us to overcome these challenges. We will consider them and their implications shortly. But first, let us see what happens when essential commodities have characteristics that prevent them from being distributed through the marketplace.

THE "COMMON PASTURE" METAPHOR

Under such circumstances, society has little choice but to regard these commodities as COMMON GOODS. This means that they are collectively owned by society as a whole rather than any of its individual members, and responsibility for them is (usually) assigned to society's government.

This has been the standard pattern for metropolitan roadway systems in the United States. They are built and maintained by a local or county (or occasionally state) government, which usually funds their costs out of general revenues. This is often supplemented by "user taxes" (i.e., taxes on the purchase of motor-vehicle fuel—which implies in theory that motorists help pay for the roadways they use based on how much use they make of them).

But even in cases where a roadway system is fully supported by fuel taxes, there remains a total disconnect in the minds of motorists between the act of using roadways and the act of paying for them. This is radically different from the case of commodities that are distributed through the marketplace, where a consumer must directly buy some quantity of a commodity before being able to use it.

The result is an instinctive sense among travelers that "roadways are free." This mistaken view greatly complicates the whole complex of issues relating to how much roadway capacity a society decides that it needs and how that capacity should be managed.

Transportation guru Joseph F. Coughlin explored this in detail in a paper titled "The Tragedy of the Concrete Commons: Defining Traffic Congestion as a Public Problem." He illustrates the basic problem by using Garrett Hardin's COMMON PASTURE metaphor, in which a society has a publicly owned pasture where local farmers are free to graze their cows without having to directly pay any user charges. Under these circumstances, each farmer seeks to graze as many cows as he can on the pasture because each cow he adds will increase his milk production with no additional feeding cost. So all farmers continue adding more cows.

This works well only so long as the total number of cows being grazed remains within the pasture's feeding capacity. But once that limit is exceeded, the viability of the pasture for grazing purposes begins to break down as its grass wears out and produces less nourishment per cow.

Since each farmer now finds that his cows are producing less milk for him to sell, his logical response is to buy still more cows and add them to the overused pasture. When all the local farmers do this, the result can only be an increasingly dysfunctional pasture and declining milk production for everyone. In Hardin's words: "Each man is locked into a system that compels him to increase his herd without limit—in a world that is limited. Ruin is the destination towards which all men rush, each pursuing his own best interest in a society that believes in the freedom of the common."

In simple terms, the problem of the overused pasture seems to be rooted primarily in the supply/demand dynamics of classical microeconomics even though the pasture's feeding capability is not distributed to farmers through a conventional marketplace. But this problem also has sociological overtones because the farmers are acting within the social context of their community. And since the pasture is publicly owned, the problem of its overuse raises political science issues as well.

The particular "commons" with which Coughlin is concerned in his paper is the nation's roadway system—especially those portions serving large metropolitan regions where traffic congestion is severe. Using terms from his pasture metaphor, he summarizes generally available data that describe the problem and lays out the various solutions being debated to address it.

For example, he notes that some people—often motorists (or farmers) who believe that their livelihoods depend on free use of the commons—propose expanding it to support larger herds (of cars or cows). Such people believe that the purpose of the commons is to serve the community's economy, so its size should keep pace with economic growth. And since the commons is publicly owned, the cost of expanding it should be paid for out of general tax revenues so that commons users can continue to obtain its benefits without having to pay for them directly. Coughlin cites Houston as an example of a metropolitan region that has followed this "expand the commons" approach to deal with traffic congestion.

At the other end of the judgment spectrum are those people who insist the real problem is not too little supply (of roadway lanes or grass) but too much demand (by motorists or farmers). They argue that the time has come to "think green" about the future of public commons in the context of the overall environment. Roadway systems (grazing pastures) are too land-hungry, and environmentally destructive in other ways. Increasing their capacity will only induce further overuse and therefore add to the environmental damage they cause. We should begin shifting to more sustainable ways of managing our communities that avoid the need for more roadways (grazing pastures). Coughlin mentions the San Francisco Bay Area as an example of a metropolitan region that has sought to address traffic congestion by investing in public transportation and various **DEMAND MANAGEMENT** techniques designed to reduce auto dependency—though the later imply tailoring our economies to transportation system capacity rather than vice versa.

As the admittedly labored pairings of motorists/farmers and automobiles/cows in the proceeding paragraphs suggest, Coughlin's grazing pasture metaphor appears to go a long way towards illuminating a broad range of socio-economic questions about why traffic congestion afflicts so many U.S. metropolitan regions.

- It illustrates the inevitable tendency towards overuse of common goods that are perceived to be "free."
- It explains why this tendency leads to a condition where supply can never really catch up with demand.
- It describes how the widespread availability of free public goods can significantly influence the underlying economics of many private activities that make use of them.
- It demonstrates the ease with which an entire society can become locked into behavioral patterns that may turn out to be counterproductive in the long run.

The obvious solution to the pasture problem is to begin charging farmers grazing fees of so much per hour per cow. This immediately confronts farmers with the need to make a series of business judgments about how much to spend feeding their cows grass in order to maximize their revenues from milk production, whether to feed them other grains instead, whether to convert their herds to higher yielding cows that have special feeding needs, and so on. When all forms of cattle feed are distributed through the marketplace at prices that reflect supply and demand, the business of milk production becomes a more rational undertaking.

The same realities hold true for metropolitan roadway systems. If it becomes possible to directly charge motorists for roadway use at so much per mile of travel, then the economics of building and managing roadway changes dramatically—and for the better.

The Movie House Metaphor

To help appreciate the implications of all this, let us consider the case of a multiplex movie house containing five separate theaters that is located in a metropolitan shopping mall with lots of foot traffic.

The whole basis for operating such a movie house is that customers must first buy tickets at the box office before they can enter any of the theaters to enjoy a movie. At any given time, the movie house manager can maximize his revenues if he can fill every single seat in his five theaters with moviegoers, with each patron having paid the highest ticket price that he considers reasonable to see a particular movie. While this rarely occurs in practice, it nonetheless remains the goal that a smart movie house manager strives to achieve. His principal tool for doing so is the price he charges for movie tickets.

The demand for access to movie seats is never constant. On any given day and at any given hour, there are a certain number of people who may be willing to buy tickets to see movies—if the price is sufficiently attractive. Because the size of this potential customer base is always changing, the smart movie house manager is going to price his tickets accordingly.

- On any given day, demand for access to movie seats is higher during the evening hours and lower in the afternoon. Therefore, the movie house manager will charge higher ticket prices in the evening than in the afternoon. Because his marginal cost to accommodate each additional moviegoer is zero during the afternoon when his theaters are largely empty, any money he collects by selling each additional ticket, no matter how low the price, is pure profit.
- On weekends and holidays, demand for access for movie seats is much higher than on normal weekdays. So the movie house manager will set evening and afternoon ticket prices at much higher levels on weekends and holidays than he does on normal weekdays.

- Not all movies are equally popular. Therefore, the movie house manager may find on any given weekend evening that one or two of his theaters showing highly popular movies have long lines of impatient patrons waiting outside for seats to become free, while his other theaters showing less popular movies have no lines and empty seats inside. Because the presence of lines can discourage some potential moviegoers, the manager can reduce their length by using lower ticket prices to attract at least some of those waiting in lines to his other theaters with empty seats inside. At the same time, he can raise ticket prices for the theaters showing the most popular movies to further encourage those waiting in line to shift to his other theaters where they can enjoy immediate seating. In so doing, the manager is maximizing the number of occupied seats in all five theaters, and is also coming closer to assuring that each movie customer is paying the highest ticket price that she finds acceptable.

There is no need to belabor this metaphor any further except to point out how effectively it illustrates the economic concept of PRICE RATIONING. This is a basic feature of free markets and is regarded as the most efficient way to allocate those goods and services (like access to movie seats—or roadway lanes) that can never be made available in unlimited quantities.

The alternative to price rationing is TIME RATIONING. This is what happens when customers have to wait in line to see a popular movie—or endure slow, stop-and-go traffic on an overcrowded metropolitan highway. It was standard practice in the former Soviet Union and its Eastern European satellites, where the prices for a wide range of consumer goods and services were set low enough for everyone to afford—but customers had to wait in line for long periods to make purchases.

Time rationing favors customers who place a low value on their time. But it penalizes those who are short of time and who might be willing to pay a higher price to avoid having to wait. A movie house manager accommodates this by offering his customers the option of paying higher ticket prices to see the most popular movie without waiting, to see movies during low-demand days like Wednesday rather than Saturday, and to attend movies in the evening rather than the afternoon.

Interestingly, the United States has traditionally used the "Leninist" practice of time rationing to control access to metropolitan roadways. With few exceptions, motorists aren't directly charged to travel on them. And most toll bridges and tunnels charge the same price during peak commuting periods as they do at three o'clock in the morning. So motorists have had to endure slow trips in stop-and-go traffic if they wish to use metropolitan roadways during the most popular travel periods. They haven't been offered the option of paying a higher price to avoid slow trips.

Moviegoers have always had this option because the technology for directly charging them for access to movie seats (and varying the price according to demand) is simple and well established. But finally, practical technology is available to directly charge motorists for access to roadway lanes in congested metropolitan regions.

The Implications of Road-Pricing Technology

For our purposes here, the physical details of the various technology systems are less important than their implications for dramatically improving:

- How the U.S. manages, maintains and pays for its existing metropolitan roadway systems.
- How it can generate new funds to build new roadway capacity to accommodate future economic growth.

In simple terms, currently available technology now allows us to distribute access to metropolitan roadway capacity through the same kind of marketplace mechanism we have traditionally used to distribute access to movie seats and a host of other commodities. This means we can:

- Directly charge each motorist for access to each roadway in a metropolitan region when he accesses it without stopping, slowing or otherwise interfering with the passage of his vehicle.
- Charge each motorist according to the distance he travels on a particular roadway, by setting a price per tenth (or some other fraction) of a mile traveled.
- Charge different prices for different roadways based on their "popularity," as measured by the number of vehicles traveling on them in the course of an hour.
- Charge different prices for different types of vehicles so that trucks and other vehicles whose greater weight causes more pavement wear pay higher prices than compact cars and other small vehicles that cause less wear.

- Raise or lower these prices at frequent intervals (such as each minute) to reflect increases or decreases in the number of vehicles traveling on a roadway.
- Use such price adjustments to guarantee motorists a certain minimum average speed on any particular roadway by raising or lowering prices to maintain a target maximum number of vehicles on the roadway and therefore a target minimum average speed.
- Apply this highly responsive differential pricing approach to all roadways in a metropolitan region—including its local streets, its major avenues, its limited-access highways, even to individual lanes on its highways.

On such a roadway system, each motorist is provided with a steady stream of real-time price and service information whenever he makes a trip by means of variable message signs along roadways or onboard message screens in his vehicle. He knows what price he is being charged to use a particular roadway, what its average speed is up ahead, and what the prices and average speeds are on other roadways that offer alternate routes to his destination. This information enables him to make rational choices about his most optimal route, given his price and speed preferences at that point in time.

Intelligent use of such technology enables us to manage entire metropolitan roadway systems with much greater efficiency than has traditionally been possible. In effect, we can narrow the often-considerable gap between a roadway system's THEORETICAL CAPACITY and its FUNCTIONAL CAPACITY.

We do this by using different price levels for roadway access to distribute travel demand more efficiently—shifting trips away from crowded roadway links to less crowded links, and away from high-demand periods to periods when demand is lower. This enables the roadway system, in the aggregate, to move more vehicles per hour at higher speeds throughout the day. Apart from reducing average travel times for motorists, this can also reduce motor vehicle air pollution because higher speeds tend to minimize a vehicle's tailpipe emissions, and shorter trip times mean that a vehicle's engine spends less time per trip generating pollutants.

We can accomplish this kind of efficient, environmentally positive trip redistribution by exploiting the classic economic principle of using price to control the demand for scarce resources. And in so doing, we can provide better service for our customers.

Yes, customers. Not motorists or travelers or taxpayers, but customers. Just like people who buy movie tickets.

This distinction is very important. If we are to be successful managing metropolitan roadway systems in a more business-like fashion and charge fair prices for using them, we must develop a business manager's habit of regarding those who use roadways as ***customers*** first and foremost.

What exactly is a customer? Someone who is a willing buyer of what we have to sell at the particular price we are charging. What makes this someone a willing buyer? Her personal judgment about whether the value to her of what we are selling is at least as great as the price we are charging.

Management guru Peter Drucker once pointed out that the most important goal for any enterprise is to CREATE CUSTOMERS—with profit simply being one of the costs we have to cover (like employee wages and supplier bills) to remain financially solvent. And Drucker insisted that the goal of creating customers is just as important for governments and other public agencies as it is for business firms.

How do we create customers? By convincing people that what we are selling meets certain of their needs and therefore has value to them, and by convincing them that this value is at least as great as the price we are charging.

Assume that a motorist can use two different highways to reach his destination in a metropolitan region. One highway charges a price per tenth of a mile that translates into a five dollar travel cost for the motorist—but it also guarantees a minimum average travel speed that will enable the motorist to reach his destination in a quarter of an hour. The other highway charges a much lower price that means a one-dollar travel cost—but the motorist knows that its traffic-choked lanes mean speeds so low that he will have to spend at least half an hour reaching his destination.

If the motorist is on his way to an important business meeting and can't afford to be late, he may decide that the value of being able to reach his destination in a quarter of an hour is at least as great as the cost of using the five dollar highway. But if he is making a trip to a shopping mall to buy garden tools, he may opt to use the less expensive highway and accept the longer travel time.

In other words, roadway pricing lets us create value for motorists by offering them the option of paying for shorter travel times when they make high priority trips. Who determines the priority of their trips? The motorists themselves. Each one makes his own judgment about how important his various trips are—and how much he is willing to pay to reach his destinations faster (for which we can offer the same kind of money-back guarantee that a movie house manager offers if the projector in one of his theaters breaks down during the movie or the sound system goes dead).

Pricing access to roadways the same way that we price access to movie seats enables us to manage our metropolitan roadway systems much more efficiently. Just like a smart movie-house manager, we can use the price mechanism to distribute travel demand at any given time in the most rational manner. If we do this properly, we end up minimizing aggregate trip times for all motorists. Shorter trip times mean better service for travel customers, enable roadway systems to process more trips per hour, and generate less air pollution in the bargain.

THE REVENUE IMPLICATIONS

Obviously, roadway pricing generates revenue. In fact, if it is applied on a wide enough scale, it can generate so much revenue that an entire metropolitan roadway system is able to become fully self-supporting by every standard business accounting measure. Such a system is paid for by the people who actually use it—according to how much they use it, when they use it, and the kind of vehicles they drive when they use it. These payments can generate a stream of revenue large enough to fund the system's annual operating and maintenance costs, the regular replacement or reconstruction of its capital plant and equipment as it wears out, the construction of new roadway capacity to keep pace with economic growth, and a competitive investment return to those who supply its capital.

This means that we can finally abandon the crude, inefficient, nonmarket funding mechanisms on which the United States has had to rely thus far for roadways. We can abolish sales taxes on motor-vehicle fuel and tires. Close down land-hungry toll plazas whose fixed-rate, manual payment toll collection procedures slow traffic to a highly polluting crawl during high-demand periods. We can even set up metropolitan roadway systems as independent, self-sustaining enterprises—moving them completely out of overburdened local government budgets where they drain general fund tax revenues away from public services like police, fire protection and education. And in so doing, we can end the unhealthy financial dependence by these systems on the federal government.

The ten largest U.S. metropolitan regions generate more than 40 percent of the nation's gross domestic product (which is a major reason why their transportation systems are so important). But the federal government's antiquated Jeffersonian biases have prevented it from treating these regions as the national treasures they are. Instead, it treats them as cash cows to support federal programs that are too often at odds with the realities of an increasingly competitive global economy (like subsidies for rural farm families to run their farms as self-indulgent art colonies rather than serious businesses).

Since the flood tide of federal spending to build the Interstate Highway network back in the 1960s, the federal government's financial support for roadway systems in metropolitan regions (which generate the bulk of federal tax revenues) has been on something like drip feed. With state and local governments increasingly burdened by federal mandates that are rarely offset by increased federal dollars, the result has been a virtual halt in serious efforts to assure that metropolitan roadway capacity keeps pace with economic growth. In fact, much existing capacity has been lost as funding shortfalls starve programs to rebuild roadways that have worn out through heavy use. The so-called "innovative financing mechanisms" proposed by transportation lobbying groups and investment bankers with excess time on their hands have rarely done much more than repackage existing federal funding authorizations in beguiling smoke and mirrors gimmickry that adds no serious new money to the nation's transportation capital funding pot.

But metropolitan roadway systems that use market-oriented roadway pricing can turn themselves into independent, self-supporting enterprises that rely on the federal government to do little more than keep the peace. Their user-generated revenue streams can assure:

- State-of-the-art maintenance and modernization of roadway systems to maximize their functional capacity without draining local tax revenues from critical public services.
- Timely replacement of worn-out capital plant and equipment so that existing roadway capacity isn't lost through age and dysfunction.
- The ability to raise fresh capital in the marketplace to build long-needed expansions of roadway capacity, so that these transportation systems can finally resume the process of stimulating new economic activity in their metropolitan regions rather than constraining it.

Conclusion

Technology revolutions often proceed faster than our awareness of how best to exploit them to improve social welfare. This is the case with the technology revolution that now allows us to price roadway access through the same kind of marketplace mechanisms that work so effectively with other commodities. It originated in the need among public authorities that operate toll bridges, tunnels and turnpikes to address the rising labor costs associated with manual collection and processing of conventional tolls.

But the technology response to this need turns out to have considerably larger implications than simply controlling labor costs. Its real importance lies in how it can be exploited to create an entirely new and more effective way to manage and finance the roadway systems that play such a critical role in the economies of major U.S. metropolitan regions.

SELF-SUPPORTING ROADWAYS

Highway Robbery Is Alive and Well

It could be argued that many of today's highway-financing problems result from the federal government's chronic inability (or unwillingness) to plan sensibly for the future. All too often, the federal response to a glaring problem is to jury-rig some sort of temporary solution that appears to be workable and then assume that it amounts to a permanent fix—only to discover after a certain amount of time has passed that events and unintended consequences have overtaken the solution and created a whole new set of problems. Then, of course, the process requires the development of a new package of solutions, which are also jury-rigged and contain hidden time bombs that will blow them apart at some point in the future.

An obvious example is the 1956 federal decision to finance the interstate highway system on a pay-as-you-build basis with revenue from federal taxes on motor vehicle fuels. This approach was contrary to the recommendation of the committee, chaired by General Lucius D. Clay, which President Eisenhower had established to advise on issues relating to the interstate system. The Clay Committee recommended that the federal government issue special purpose bonds to finance the system and support these bonds with dedicated revenues from fuel taxes so that future generations of motorists would be paying for their share of the system's use.

For such bonds to be acceptable to the capital markets, they would have had to be accompanied by specific legal covenants designed to insure their viability, the viability of the fuel-tax revenue stream supporting them, and the viability of their underlying assets (i.e. the interstate highways themselves). A consistent pattern of legal rulings from the

---

Adapted from Joseph M. Giglio, "Highway Robbery Is Alive and Well" (speech, IBTTA Transportation Finance Summit, Washington, DC, March 3–5, 2004).

U.S. Supreme Court on down has established the common-knowledge principle that bond covenants are sacred. So long as bonds remain outstanding, their covenants cannot be violated. Not by subsequent government regulations. Not by subsequent lawsuits. Not by subsequent court decisions. Not by subsequent acts of any state legislature or Congress itself. And not even by subsequent amendments to the Constitution itself.

Therefore, if the Clay Committee's bond-financing recommendation had been adopted, none of the following could have occurred or could even have been debated in Congress:

- Diverting federal fuel-tax revenues to support non-highway spending.
- Artificially suppressing federal highway spending to build up surpluses of fuel-tax revenues in the Highway Trust Fund that the U.S. Treasury could "borrow" to fund other government operations.
- Failing to index fuel-tax rates as needed to provide sufficient revenues for appropriate levels of ongoing maintenance, timely reconstruction and new capacity for the interstate highway system.

But the Clay Committee's recommendation ran afoul of certain powerful conservative voices in Congress. Some of them were concerned about the "extra cost of bond interest" because they were unable to understand that the present value of such interest costs (which is what counts in measuring their true impact) would be zero over the life of the bonds. Others were opposed to any federal debt that would not be subject to congressional scrutiny and approval.

So the end result was a "compromise." No bonds would be issued to fund the interstate highway system. Instead, construction would be financed on a pay-as-you-build basis with fuel-tax revenues that would be deposited in the Highway Trust Fund. Annual appropriations from the Trust Fund would then be determined by proposals in administration budgets and congressional approval.

This compromise held together long enough for most of the interstate highway system to get built. It began to unravel as it was overtaken by events (improvements in motor-vehicle fuel efficiency that reduced fuel consumption per mile of travel) and unintended consequences (diversion of much fuel-tax revenue into Highway Trust Fund surpluses that were borrowed for non-highway purposes). This brought the U.S. to the financing predicament that it now faces. Today, the federal government does not have enough money to expand highway capacity in the face of rising travel demand generated by a growing economy; reconstruct critical highway links when they reach the end of their useful lives; restore highways that deteriorated too rapidly because of inadequate maintenance; and even maintain existing highways sufficiently to prevent them from deteriorating further. All of this leaves the nation's highway system living on borrowed time.

In the last analysis, there are only two sound ways to finance highways:

1. With general tax revenues collected from the public at large. This assumes that highways are no different from other public goods (such as police and fire protection), whose benefits extend beyond their immediate value to those who make use of them, to ultimately impact society as a whole in various and complex ways.
2. With user-charge revenues collected from motorists traveling on them. This assumes that highways are no different from other marketplace goods like electric power and first-class mail service, whose benefits can be fairly priced to those wishing to make use of them and who will ultimately pass on some portion of these costs to the rest of society through the marketplace.

But Americans managed to convince themselves that there was a third way when it came to highways, one that attempted to combine features of the first two. It involved taxing only motorists (not society at large) by adding what amounted to a surcharge to the prices they paid for motor-vehicle fuels and using surcharge revenues to fund highways.

This is equivalent to a movie theater that doesn't charge admission to moviegoers. Instead, it adds surcharges to the prices of the popcorn, candy and soda they purchase in the lobby. With these surcharge revenues, movie theaters would pay rental fees to the distributors of the movies it shows, fund the cost of other theater operations and provide the necessary return on the capital invested to build the theater. While this sounds too outlandish ever to be taken seriously by the movie business, it obviously is not by the highway business.

Thus, surcharges on the prices of secondary goods associated with highway use became the main source of revenue to support American highways (with a few toll roads scattered here and there to highlight the difference). Its greatest success occurred during the 1960s when most of the interstate system was built.

But over time, the inherent disconnect between surcharge revenue from secondary goods and rising highway-funding needs due to the increasing use by motorists ballooned into a serious problem. Surcharge revenue became constrained by consumer pressure for more fuel-efficient vehicles in the wake of OPEC price increases during the 1970s (akin to weight-conscious moviegoers reducing their consumption of popcorn, candy and soda in the movie theater). And federal control over fuel-surcharge revenue caused much of it to be diverted to non-highway uses (like the movie-theater manager skimming off surcharge revenue from popcorn, candy and soda sales to pay rent on an apartment for the busty blond usherette who has become his mistress). This eventually led to a gap between highway-spending needs and available funds, a gap that became increasingly worrisome during the 1990s.

The result was that the political hills became alive with the sound of heated debates over proposals to address this gap. Most often, this took the form of creative financial engineering tools to leverage the buying power of available funds, rather than attempting to increase the funds themselves. As with many such debates, much of the heat surrounding these "innovative financing tools" was fueled by imaginative speculation rather than serious analysis of empirical data.

The ultimate goal is not to supplement the fuel tax, but to supplant it. Yes, additional financing techniques and tools are important. But to remedy the underinvestment gap, we need to focus on current and future roadway pricing technologies as enablers of positive change. The physical details of the various technologies are less important than their implications for dramatically improving how we manage, maintain and finance highway capacity, and how they can generate net new funds to add highway capacity to accommodate future economic growth.

The key phrase here is "net new funds." This means more than using financing techniques simply to shift monies from one time period to another. It means developing new revenue structures to replace indirect user charges. Truly innovative finance adds net new resources to transportation.

In simple terms, technology now allows us to distribute access to highway capacity through the same kind of marketplace mechanisms that we have traditionally used to distribute access to movie seats and a host of other commodities. These technologies represent an opportunity to finally abandon the crude and inefficient non-market funding mechanisms on which the United States has relied to finance its highways. They present a chance to gradually jettison the "Leninist" practice of time rationing to control access to highways. Equally important, they enable you to regard those who use highways as customers first and foremost. Yes, customers—not motorists or travelers or taxpayers, but customers.

The distinction is key. If you are going to be successful in managing highway assets and charge market prices, you need to regard those who use highway systems as customers. By customers, I mean those who are willing to buy what you have to sell at the particular price you are charging. What makes someone a willing buyer? Their judgment as to whether the value of what you are selling exceeds the price you are charging.

In the highway business, you do this by improving the quality of what you are selling. Exploiting current technologies, such as GPS systems, facilitates this value proposition. Technology is the enabler here. In sum, you have the technology to apply to transportation what is essentially performance-based pricing. Of course, these technologies also have major implications for safety, security and congestion.

I don't know if the path I am suggesting is simply paved with good intentions or gold (or bonds, as the case may be). But I'm confident we need interaction and greater alignment between finance, technology, customer service and the management of our highway enterprise.

SELF-SUPPORTING ROADWAYS

You Have to Pay If You Want to Play

When it comes to the outlook for financing necessary improvements to the nation's surface transportation systems, a stark reality is emerging. The federal government simply is not going to be able to meet its obligations in this area.

Never mind what kind of pipe-dream authorization bill Congress and the White House manage to cobble together in the next year or so. Appropriations in the years that follow are what count. And these annual appropriations are going to be smacked in the face by huge federal budget deficits as far as the eye can see from such problems as:

- The escalating costs of supporting U.S. military activities in the petroleum-critical Middle East.
- The struggles to put together effective homeland-security systems without turning the country into a police state.
- The increasing burdens of keeping Social Security and Medicare more or less solvent as they become inundated by waves of retiring Baby Boomers.
- The growing need to find taxpayer support to prevent private pension and medical benefit systems from overwhelming a disturbing number of our most important corporations.

According to the American Association of State Highway and Transportation Officials, the United States will face a $1 trillion backlog of unmet needs during the first quarter of the 21st century simply to prevent the badly deteriorated condition of the nation's highway system from

---

Adapted from Joseph M. Giglio, "You Have to Pay If You Want to Play" (presentation delivered to the Transportation Research Board, Washington, DC, January 10, 2005).

getting any worse. After adding in the costs of improving the system to overcome the effects of past underspending and to keep pace with growing travel demand, this funding need rises to almost $2 trillion. Figure 1 below shows the projected cumulative backlog of transportation needs through 2025 and the anticipated gaps assuming various estimates.

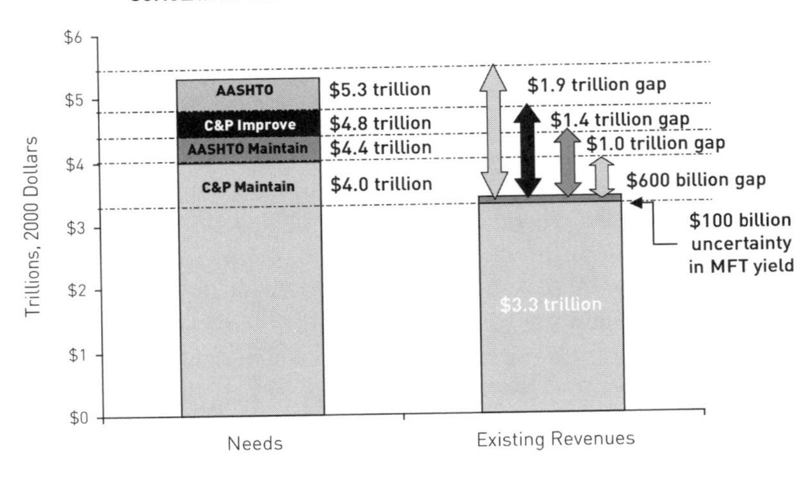

Source: Cambridge Systematics based on FHWA and AASHTO data.

BACK TO THE STATES

Under normal circumstances, needs of this magnitude would call for a major financing commitment by the federal government to supplement the ongoing efforts by hard-pressed state and local governments. But today's circumstances aren't what we can call normal and we can't reasonably expect the necessary federal funding to materialize. This means that an increasing portion of this funding burden will have to be carried by the states.

On its face, this sounds like a new twist on the devolution concept. Devolution is a fancy academic name for a process under which the feds stick the states with expensive new responsibilities while still collecting the lion's share of every total tax dollar each state generates. An obvious example is homeland security, where the states with the greatest anti-terror needs receive the fewest dollars per capita of federal aid. In the case of transportation, the feds are using the budget appropriation process to force the states to assume what has previously been federal funding responsibility for existing programs, rather than passing laws mandating new programs that the states have to fund. But devolution by any other means is still a wilted rose.

The states can't simply plead poverty and unfairness as excuses for trying to back away from assuming a greater share of what was formerly a federal responsibility, because transportation provides essential support for the economy of each state and the nation as a whole. We have to remember that economic activity generates travel demand. And this demand must be met with transportation services of respectable quality if economic activity is to grow fast enough to provide decent living standards for a growing population.

This is why proposals to emphasize something called Transportation Demand Management represent dead-end thinking. Such proposals implicitly assume that we should tailor our economy to the size and capacity of our transportation systems, and this is 180 degrees wrong. Transportation systems exist to support the economy, not vice versa. So the states are going to have to step up to the plate to ensure they have the transportation systems needed to grow their economies.

The Power of Technology

Fortunately, the states have an important new ace up their sleeves when it comes to shouldering the burden of providing the nation with the modern roadway system it needs in order to remain competitive in the global economy. If they play this ace wisely, it can even enable them to revolutionize in some very positive and sensible ways how the nation's surface transportation systems are financed and managed.

This ace happens to be a comparatively new, remarkably simple and reliable technology that allows states to collect highway tolls from motorists without requiring them to stop and fumble for change or tokens at land-hungry toll plazas. Those of you who are from New York or New Jersey are already familiar with this technology, where it goes under the name of E-ZPass. Indeed, this technology encompasses 22 different state agencies in 11 states. The same is true for those of you who are from Southern California, Texas and a few other places around the country where this technology already operates successfully.

The primary advantage of this technology is that it enables states to charge motorists for access to highway lanes and to collect fees for each trip based on the number of miles traveled. This means we can provide highway travel services through the same kind of market-based price system we use to provide most other consumer goods and services. Motorists pay for the travel services they use. And each one makes her own judgments about the trade-off at any given time between the price she is being charged for highway access and the value to her of the service that he is buying.

We can go further and price highway access according to the type of vehicle the motorist is driving—charging him a higher price per mile if he is driving a heavy truck whose weight causes more pavement wear and a lower price per mile if he is driving a lightweight compact car. We can even vary the price according to demand, just as we do in movie theaters where the admission price is higher on Saturday nights and lower on weekday afternoons. And we can offer the motorist a money-back guarantee that his average travel speed on a particular highway won't fall below a certain miles-per-hour standard. This would provide an economic incentive for the highway operator to maintain the pavement surface in top condition at all times, to quickly clear away lane blockages from accidents and disabled vehicles, and to make use of the most effective traffic-control technology to keep the flow of vehicles moving at a reasonable speed.

Finally, we can use intelligent pricing to manage our roadway systems more efficiently, especially in urban areas where travel demand is high. By changing the price per mile on a real-time basis in response to traffic volumes, we can encourage some traffic to shift away from high-demand highways to routes where demand is lower, and from time periods when demand is high on all routes to periods when demand is more modest. The greater efficiency this promotes can enable roadway systems to move more vehicles per day at higher speeds without incurring the expense, long construction times and environmental problems of building additional highway lanes. For the first time, we can raise the functional capacity of our roadway systems much closer to their theoretical capacity.

THE ROAD TO SELF SUFFICIENCY

But perhaps the most practical advantage of this modern approach to highway pricing for the states that adopt it is that they can make their roadway systems fully self-supporting from user charges. Liberating them forever from any further reliance on fuel taxes, old-fashioned toll plazas, annual appropriations from revenue-short state and local government budgets, and constant begging for alms from an increasingly parsimonious federal government. Functioning instead as sound commercial enterprises under state guidance. Able to fund better operations, better maintenance, better traffic control facilities and long overdue capacity enhancements from the revenues they generate. Purchasing the best the private sector has to offer in order to provide the best service to their motorist customers.

States that adopt this approach will have to be guided by an effective strategy in order to generate maximum dividends in the shortest time. The core of this strategy is to concentrate on two functional areas. One is the roadway systems of their metropolitan regions, where most travel takes place. The second is the travel corridors that link these metropolitan regions to each other and to the rest of the country.

Ten large metropolitan regions in the United States generate some 40 percent of its gross domestic product even though they contain only 31 percent of its population. This means that, taken together, their per capita GDP is about 28 percent greater than the nation's per capita GDP. This is why we should recognize them as national treasures and the key to future national prosperity, and why the states that encompass them should give them priority.

Per Capita GDP in 16 Major U.S. Metro Regions

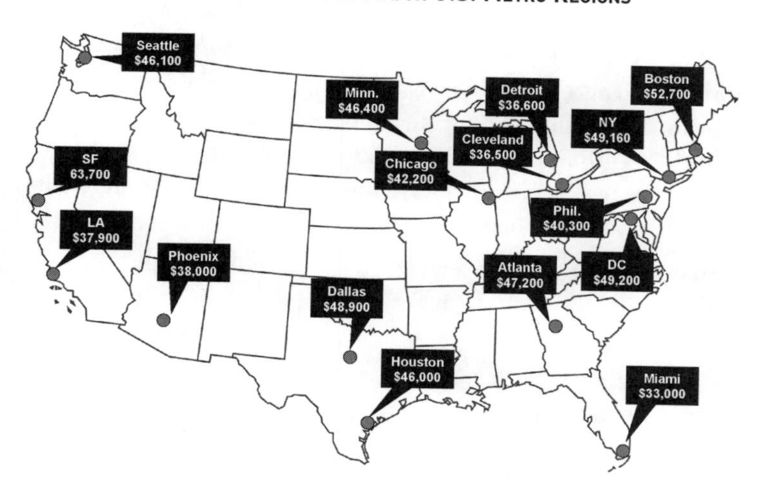

Source: U.S. Census Bureau

In technical terms, each of these ten regions is formally known as a Consolidated Metropolitan Statistical Area. While each region is centered on a well-known major city, its boundaries extend far beyond the city limits. Metro Washington, for example, includes the District of Columbia, plus a large portion of northern Virginia, plus those portions of Maryland that encompass Baltimore and Hagerstown and their suburbs. Five of these ten metro regions are each contained within single states. The other five include portions of several states, which suggests the need for a new emphasis on cooperation between states.

Following is the list of the top ten metro regions, ranked by their gross domestic product.

1. **Metro New York**, whose GDP is about the size of Italy's. It spreads across three states and encompasses New York City, Long Island, the Mid-Hudson Valley, northern New Jersey and southeastern Connecticut.

2. **Metro Los Angles**, whose GDP is about the size of Mexico's. It is entirely contained within the state of California and encompasses the cities of Los Angeles and Long Beach and their suburbs, plus Orange County, Riverside County, San Bernardino County and Ventura County.

3. **Metro San Francisco**, whose GDP is about the size of South Korea's. It is also entirely contained within the state of California and encompasses the counties of San Francisco, San Mateo, Marin, Alameda, Contra Costa, Santa Clara, Santa Cruz, Sonoma, Napa and Solano.

4. **Metro Chicago**, whose GDP is about the size of the Netherlands'. It spreads across three states and encompasses the counties of Cook, DeKalb, DuPage, Grundy, Kane, Kendall, Lake, McHenry, Kankakee and Will in Illinois; the counties of Lake and Porter in Indiana; and the county of Kenosha in Wisconsin.

5. **Metro Washington**, whose GDP is also about the size of the Netherlands'. As noted earlier, it spreads across three states and encompasses the District of Columbia, northern Virginia and much of Maryland.

6. **Metro Boston**, whose GDP is about the size of the Russian Federation's. It spreads across four states and encompasses the City of Boston, the northern suburbs running up through New Hampshire into southern Maine, the southern suburbs down to Fall River and

New Bedford, and the western suburbs running out to Worcester and down into Connecticut's Windham county.

7. **Metro Dallas**, whose GDP is about the size of Switzerland's. It is entirely contained within the state of Texas and encompasses the cities of Dallas and Fort Worth and their suburbs.
8. **Metro Philadelphia**, whose GDP is also about the size of Switzerland's. It spreads across three states and encompasses the city of Philadelphia and its suburbs in Pennsylvania, extending as far east as Atlantic City in New Jersey and as far south as Wilmington in Delaware.
9. **Metro Houston**, whose GDP is about the size of Sweden's. It is entirely contained within the state of Texas and encompasses the city of Houston and its suburbs, extending as far south as Galveston.
10. **Metro Detroit**, whose GDP is about the size of Austria's. It is entirely contained within the state of Michigan and encompasses the cities of Detroit, Ann Arbor, and Flint and their suburbs.

The chart below provides a summary of the top ten U.S. metropolitan regions ranked by gross domestic product.

Top Ten U.S. Metro Regions by GDP

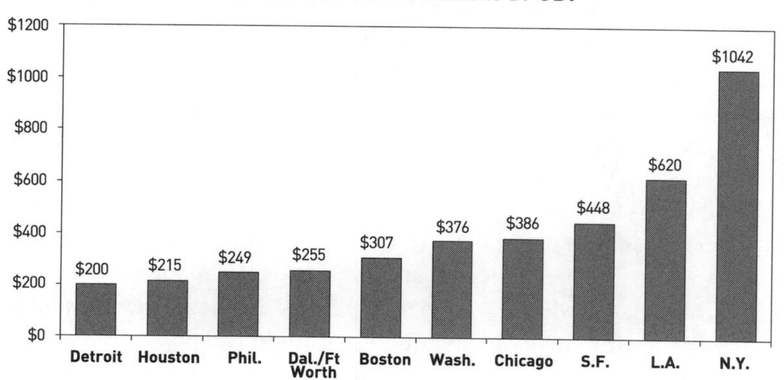

Source: U.S. Census Bureau

As it happens, metropolitan regions are also major constituents of the nation's primary travel corridors. They represent trip origin and destination points as well as key junctions along the way for through trips. So when states concentrate on making the highway systems of their metro regions self-supporting with user charges, they are more than halfway to accomplishing the same thing for their entire travel corridors. This is the practical significance of how the revolution in roadway financing can proceed.

We can understand this better by looking at Metro Boston as an example. As we just saw, it is the nation's sixth largest U.S. metropolitan region in terms of GDP. It encompasses portions of four states and is directly adjacent to a fifth state. More importantly, it is the northern terminus of the Northeast Corridor, which is the highest volume travel corridor in the United States.

Ideally, the Massachusetts Executive Office of Transportation should work out a set of common standards for highway user charges with the state transportation departments of New Hampshire, Maine, and Connecticut. These standards would presumably cover such issues as what technology to use and what rates per mile to charge. But this ideal may not actually be necessary in practice.

For example, New Hampshire may decide to adopt highway user charges but prefer a different rate structure, while Maine may decide to forego user charges entirely for the time being and remain dependent on federal financial aid. But none of this really matters. All that's necessary is for Massachusetts to make available to all motorists in New Hampshire and Maine its standard transponder for toll collection.

Interestingly, the fact that states do not have to agree up front on common standards could actually make it easier for them to achieve these standards in the long run. But that's the American way.

Where does this leave the federal government under such a financing system? That's an interesting question.

If you subscribe to the principle that "you have to pay if you want to play," you can argue that the feds should have no role at all in states that elect to make their roadway systems fully self-supporting with user charges. The feds wouldn't collect fuel taxes in these states, wouldn't provide them with any roadway funding, wouldn't impose national standards on their roadways, wouldn't try to tell them what roadways to build and where. The federal government would be out of the roadway transportation picture completely in these states. Instead, its role would be confined to those states that choose not to adopt user charges and prefer to continue under the present federal/state arrangement.

Which is just fine. No state should be forced into adopting anything it doesn't think meets its needs. There are a number of states whose rural nature, large land areas, low population density, lack of significant metropolitan regions, below-average per capita GDP, and ability to bargain in Congress for favorable treatment have given them a long history of dependence on receiving more financial support from the federal government than they contribute in tax revenues. Retaining some version of the present federal/state arrangement and preserving the federal fuel tax may be the best option for them. But encouraging more states to make their roadway systems self-supporting could be an effective way for the federal government to reduce the growth of overall federal aid to states in the face of its severe budget deficits.

The Federal Government's Role

Some of you may wonder how all this squares with the constitutional requirement that the federal government promote interstate commerce. If we interpret this requirement to mean only direct federal aid programs and policies, there are obvious problems.

But such an interpretation may be much too narrow. Promoting interstate commerce need not be limited just to hands-on federal involvement. It can also include more passive and subtle federal encouragement of state initiatives and the provision of greater opportunities for them to allow free-market economics and private-sector involvement to play a larger and more positive role.

For example, California is rapidly becoming the national standards setter for motor-vehicle emissions. Its severe air pollution problems have required it to act more aggressively than the federal government in setting more rigorous limits on new vehicle-exhaust emissions. And the immense size of its new-vehicle market has increasingly caused the automobile industry to adopt California standards for all new vehicles sold in the United States, especially as more states choose to piggyback on California's initiatives in this area. The result is that we are rapidly approaching the point where federal emissions standards are becoming irrelevant, and could even be scrapped.

Given these circumstances, it doesn't require any great leap of imagination to foresee state initiatives becoming the de facto national standards in such areas as toll-collection technology—even to the point where sophisticated on-board toll collection and monitoring equipment become a fact of life in all new vehicles sold in the United States far sooner than might otherwise be the case.

In theory, the federal government could retain at least a national policy guidance role in surface transportation, and the specifications of this role could take a number of different forms. With the nation at war—probably for years to come—states can presumably benefit from federal guidance in the area of transportation security, since 9/11 demonstrated that travel vehicles can be turned into highly destructive missiles. The federal presence can also be beneficial as a referee, as an information clearinghouse, as a facilitator of compatibility among various toll collection systems and as a sponsor of transportation research and development. All this emphasizes enlightened thinking on the part of the feds rather than dispensing lots of dollars.

Unfortunately, enlightened thinking is in short supply in Washington these days. The feds' ability to deal with serious transportation policy questions has diminished significantly from its 1950s, take-charge, 90-percent funding and policy responsibility for building the Interstate Highway System. Increasingly, new ideas in Washington have focused on saving money by emphasizing "good decision making at the state and local levels." That's what's really behind the frequently repeated federal mantra of "We can't build our way out of congestion." (Of course we can, especially if we take a wide-angle shot and consider other modes as well. And we must. But it will cost money that the feds don't want to provide.)

For evidence of this, we need look no further than the present discussions about transportation reauthorization going on in Congress and the White House. These discussions are entirely budget driven. Their focus is on the nuts and bolts of how the feds can limit future spending. The policy issues of how the United States can develop the transportation capability it needs to remain competitive in the 21st century are never addressed. The states must fill this critical vacuum.

THE NEW INTERMODALISM

We must keep one final point in mind as we think about having states adopt motorist user charges to make roadway systems self-supporting and independent. In practical terms, user charges constitute a new revenue source for surface transportation as a whole. Perhaps this is the last new revenue source available unless we are prepared to commit a much larger portion of general tax revenues to transportation, which seems unlikely.

Note the deliberate choice of the words "new revenue source for surface transportation as a whole." In other words, we're talking about more than just highways. Because motorist user charges have the potential for enabling each state to address the larger problem of truly integrating all of its surface transportation networks as a full-fledged intermodal system rather than remaining an unrelated group of separate modes.

Many of you will see the obvious application of this for the issue of public transportation in major metropolitan regions. Public transportation may generate enormous economic benefits in these regions. But there are few practical ways to capture a fair share of these benefits to provide the revenues that public transportation networks need. Existing combinations of rider fares, sales taxes, real-estate taxes and even payroll taxes all fall short. So we continue to struggle with ways to force metropolitan roadway systems to accommodate trip volumes that might more sensibly (and even less expensively) be handled by public transportation. Highway user charges can provide a new revenue source to change this.

Less obvious but perhaps equally important is the issue of goods movement—not only within metropolitan regions but also within interregional travel corridors. As trip volumes for goods movement increases with the growth of economic activity, our existing and badly fragmented surface transportation system inevitably translates these increases into more truck trips on roadways—much to the dismay of automobile drivers—even when many of these trips could be handled more efficiently and less expensively by rail.

But the nation's rail-freight network has greatly diminished in size and capacity in recent years as the privately owned freight-railroad companies have scrapped thousands of miles of railroad track in order to reduce maintenance costs. This has continued to the point where, for example, the Union Pacific railroad company had to send United Parcel Service a Dear John letter last March terminating the arrangement the two had worked out under which Union Pacific would operate special transcontinental trains of UPS parcel containers. Union Pacific had to do this because rising commodity freight volume had left it with no track capacity to handle UPS trains. So all these containers now have to move by truck on highways.

The issue here is who should pay for the cost of the additional track capacity needed to move a greater volume of freight by rail. If the alternative is to crowd the nation's highways with more truck trips, doesn't it make sense for states to use some of their motorist user charge revenue to help cover the cost of assuring more track capacity for goods movement rather than spending it only on more roadway capacity for trucks? This is another example of how the intelligent use of highway user charges by states can help us move towards a more integrated and intermodally efficient surface transportation system for people and goods.

But wait. Aren't we talking about cross-subsidization now? And isn't that a dirty word? Especially among those academic theorists in university B-schools and economics departments who insist that every product line must pay its own way in a narrow cost-accounting sense?

Well, sure. But that isn't the real world. The real world is the successful multi-product corporation that recognizes how different products contribute in different ways to its bottom line. Like the pharmaceutical company that continues to market a traditional drug even though the fact that its patents have long since expired has turned it into a low or even negative margin product. Why? Because it knows that lots of people have come to depend on this drug and the company that produces it, and that these people represent a ready-made customer base for the next high-margin pharmaceutical breakthrough due out next year. So the old-line product that may not even pay its way in the strict cost-accounting sense contributes significantly to the powerhouse profits of new products in ways that totally scramble the overly rigid assumptions of traditional cost accounting and portfolio management.

CONCLUSION

The outlook for federal surface-transportation-funding reauthorization may seem to be a bleak one right now. But hopefully, it will motivate at least some states to find clever ways to exploit the opportunities presented by motorist user charges. In which case, it can turn out to be a long-needed breakthrough in giving the nation's most economically productive regions the kind of integrated surface transportation systems they need.

I suppose a lot of this may seem distinctly Jeffersonian to those of you who have read American history. Thomas Jefferson was born as the rich and pampered the son of a plantation owner and never had to confront the ice-water-in-the-face reality of having to earn a living. So he was free to spin his fairyland visions of an agrarian America full of independent farmers, where government initiatives were largely left to the states rather than to a strong central government.

But it seems to me that the discussion here actually draws more on the traditions of Alexander Hamilton. Hamilton was born poor and had to claw his way up the financial ladder on the strength of his ever-pragmatic intelligence and his ability to see workable connections between vision and reality. The key is pragmatism. States that adopt highway user charges to generate a new revenue stream for transportation will be responding pragmatically to a federal abdication of funding responsibility. Which they can turn into much better and more integrated surface-transportation systems. Very Hamiltonian in spirit.

SELF-SUPPORTING ROADWAYS

Private Toll Roads: Is the Glass One-Quarter Full or Three-Quarters Empty?

Those of us who follow government and public affairs have been hearing a great deal about how privately financed and operated toll roads may be the best solution to the nation's chronic underinvestment in its highway system. Here are some background facts on "privatization" and questions to ask when evaluating private toll-road proposals.

The Investment Problem

In recent decades, the United States has been investing too little in the roads and bridges of its highway network. Right now, we should be spending at least $20 billion more per year simply to keep these essential transportation facilities from deteriorating any further. To fund the upgrading, modernization and expansion needed to support a growing economy would require a $40 billion per year increase in what we're currently spending.

A key reason for this lack of adequate investment is the magical transformation of what were once called "dedicated transportation user charges" (such as motor-vehicle fuel taxes) into sources of general tax revenue. The federal government has been reducing its budget deficit by holding down infrastructure outlays from the Highway Trust Fund so that the resulting surpluses of annual user-charge revenues can be used to help cover non-transportation expenses. Many state governments are doing the same thing.

Meanwhile, the world's financial markets are awash with private capital looking for economically sound infrastructure projects in which

---

Adapted from Joseph M. Giglio, "Private Toll Roads: Is the Glass One-Quarter Full or Three-Quarters Empty?" Nieman Reports *(The Journalism Quarterly at Harvard University) 51, no. 4 (Winter 1997): 25–28.*

to invest. Several multibillion-dollar infrastructure funds have already been assembled and more are being developed. This has led to the concept of having private firms build self-supporting toll roads that can tap these infrastructure funds for construction dollars.

This concept seems to be working effectively in France, Spain, Italy, Portugal, Indonesia and even China. But a quick scan of North America shows a landscape littered with the dry bones of failure and a rapidly aging cadre of privatization advocates. The smart money behind private infrastructure funds seems to be finding opportunities in every country except the United States.

WHY HAS THE U.S. FALLEN BEHIND?

Full-fledged privatization of transportation facilities has largely been a failure in the United States. The few successes have come only after years of complicated negotiations and a sad history of aborted projects.

California's Assembly Bill 680 was passed in 1986 to stimulate construction of private toll roads throughout the state. Since then, negotiations have been completed with private firms for four such projects. But it wasn't until late 1995 that the first one opened to traffic.

That project was the 10-mile, four-lane, all-electronic toll road in the median of State Route 91 in Orange County. While its initial traffic volumes were less than originally projected, the growth has been encouraging and public acceptance of variable rate tolls and electronic toll collection has been good. However, no information has yet been released to show whether the road is a financial success.

The other three California projects have made little progress, largely due to state insistence on 100 percent private funding.

The experience in other states has been even more disappointing.

In Washington, six projects were selected for development under the state government's public-private partnership initiative. But a dramatic change in the political composition of the state legislature following the 1994 election resulted in anti-toll sentiment that virtually halted privatization efforts. The only project now underway is a series of park-and-ride garages in Seattle. But this project is funded entirely with public capital and is therefore not an example of true privatization. However, the Tacoma Narrows toll-bridge project could end up as a truly private project if it survives a regional referendum in late 1998.

Virginia's Dulles Greenway toll road was an ambitious privatization project that cost over $300 million and received substantial private equity (reaching as high as 50 percent of total capitalization at one point). But the highway has experienced low traffic volume and isn't generating enough toll revenue to cover its debt service. Long-term economic growth in the region it serves could eventually solve this problem, but the road's ownership may change hands several times before then. Meanwhile, the I-895 Richmond Connector Toll Road project has struggled to get past the preliminary stages and now appears to be stalled because of disagreement between the state government and private developers over how much private capital the developers should invest.

Arizona has been through three rounds of privatization attempts. Two ended in failure due to community opposition, poor economics and a lack of political support.

South Carolina's experience with highway privatization has been mixed: one complete failure due to local opposition, one apparent success (the Greenville Toll Motorway), two more still in negotiation.

Only California's State Route 91 and Virginia's Dulles Greenway can be categorized as true private toll roads, because they embody the significant private equity investment and exposure to risk and reward that characterize all private-sector undertakings. Both were conceived nearly a decade ago and neither is yet a financial success.

The other non-traditional toll roads are basically public-private partnerships, with government providing all or most of the investment capital while private firms do the building and operating. Most of the recent progress has involved projects that have little private investment and use phased construction to accelerate project completion. Their success seems to depend on:

- Having well-focused project goals, often determined in advance by government.
- Allowing the public and private partners to lead with their strengths. This usually means having government fund the project with tax-exempt debt, while the private firm takes on the phased-construction management responsibilities.
- Alternately structuring the project in ways that fall well short of true privatization and may simply involve some variation of the well-established "contracting out" concept (Virginia's Interstate Maintenance program is an example).

ELSEWHERE IN NORTH AMERICA

In Mexico, the good news is that several major private toll roads have been built. The bad news is that most have been financial failures, even if political triumphs. In general, these projects were undertaken as ways to support Mexico's construction industry. Tolls were set at high levels in the mistaken belief that this would permit rapid repayment of construction debt. But this only succeeded in discouraging motorists from using the roads. While some smaller toll roads in urban areas appear to be self-supporting, most of these projects are in the process of being refinanced around lower and more realistic toll rates.

Canada has had at least one significant success with toll projects and few outright failures. This may be due to limited promotion of privatization for its own sake. The Northumberland Straight toll bridge between Prince Edward Island and the mainland opened during the summer of 1997. Its construction was privately financed, but it receives annual operating subsidies from the Canadian government from funds that formerly subsidized the ferries that the bridge replaced.

Route 407 outside Toronto has been a major success in terms of construction efficiency. But it was never promoted as anything more than a large-scale phased-construction project. Route 104 in Nova Scotia (a truck by-pass around Halifax) may also end up as a hybrid success, with a combination of provincial government debt capital and private equity funding.

THE DIAGNOSIS

The problems faced in developing more private toll roads in the United States fall into three broad categories:

1. The public sector's access to ostensibly low-cost, tax-exempt debt, which discourages consideration of alternate financing mechanisms.
2. Poor communications between the public and private sectors. This includes a broad set of public-sector approval processes that frustrate private developers and sometimes lead to outright failure.
3. The multiple (often conflicting) objectives of private firms that can get in the way of structuring a reasonable deal for the project.

A key factor influencing project financing decisions is the access of state governments to relatively low-cost, tax-exempt debt. When toll roads have promising economic prospects, government is usually reluctant to turn the financing (much less the project itself) over to a private partner.

One effort to get around this is the increasing use of "63-20" corporations. These are not-for-profit corporations that the Internal Revenue Service allows to issue tax-exempt debt for private development. Even so, many private firms have been unwilling to make significant equity investments in toll-road projects. They much prefer to serve as project developers in return for a flat fee. In addition, the modest cost of establishing "63-20" corporations has discouraged some local governments from pursuing private-sector participation.

With very few exceptions, private developers have been unable to convince public agencies that imagination and efficient management can offset the apparent financial benefits of tax-exempt debt. Ironically, financial advisors, the public-finance arms of investment-banking houses, and other private players in the tax-exempt debt market are often the strongest advocates of maintaining the status quo.

Despite much apparent goodwill on all sides during the early stages of project development, private firms and public agencies often speak different languages, have different sets of values and follow different practices. Typical problems include:

- ***A different pace of decision-making.*** Public agencies are accustomed to a slower pace of decision-making than is usual in private firms, in part because there are more players involved. For example, environmental agencies and community groups can impose delays in the decision-making process.
- ***Government's lack of a single decision-maker.*** Even when approval has been reached within one agency, another branch of government may change the rules (as happened in the state of Washington) or even halt the approval process (as Arizona's governor did during the first round of privatization projects in that state).
- ***Government's complex procurement process.*** The traditional practice of competitive bidding is often at odds with a private firm's need to protect its proprietary ideas. Some states have been able to develop creative ways around this, as Virginia has done with its very short deadlines for bids. But all too often, private-sector mistrust in the ability of public officials to follow through on good intentions leads to a lack of serious bids (as has happened in Delaware).

- ***Unrealistic financial expectations on both sides.*** Many public officials see the private sector as a source of easy money. But they fail to appreciate the need for a fair rate of return on private equity investments because of a suspicion that "private profits rob the public." (Interestingly, this attitude seems to be less of a problem outside the United States.) At the same time, few private firms have shown much willingness to make serious equity investments in projects. As already noted, they tend to be more interested in getting paid to build toll roads than in the revenues such roads can generate.
- ***Limited support in public agencies.*** In many state transportation agencies, few officials are truly interested in private-sector involvement. Most prefer to continue doing business as usual, even if that means construction delays and higher project costs.
- ***Motorist opposition to tolls.*** This is widespread regardless of whether the toll collector is a public agency or a private firm. Despite funding shortfalls, there is often a stronger faith in the power of pork-barrel politics than in the efficiency of the marketplace.
- ***Finally, private developers can fall just as blindly in love with their projects as public-agency developers.*** A project's underlying merits need continuing re-examination to determine whether changes in its scope are in order. A scaled-back Dulles Greenway could have brought costs more in line with actual travel demand. Instead of simply assuming that motorists "must" want to use the road as originally planned, more up-front market research could have revealed demand levels that dictated a less ambitious road.

THE FUTURE

The first round of private toll roads in the United States has fared poorly. This has led to renewed emphasis on developing conventional public-private partnerships for public-agency toll roads, often utilizing phased construction. But real opportunities still remain for truly private toll roads if three points are kept in mind:

- ***Innovative project financing is still alive and well.*** The assumption that conventional tax-exempt financing by government produces interest cost savings shouldn't obscure the larger financial benefits offered by other options. For example, federal law now permits state loans at below-market interest rates to private toll-road developers.

- ***The layered look is in.*** Private developers need to learn how to take full advantage of state infrastructure banks and other innovative public sector financial tools. Also, there is no need to rely solely on tolls to support a privately developed road. It is possible to take advantage of the increased property values and other economic benefits produced by a new highway to build a financing package around several revenue sources. Route E-470 outside Denver and the San Joaquin Toll Road in California's Orange County rely on as many as five different revenue sources.
- ***Just do it.*** Private and public sector players too often have multiple objectives, all of which would be better served by a single-minded focus on getting the project built.

Four Questions to Ask

Evaluating the viability of a new private toll-road proposal is no easy task. But here are four questions to consider.

*1. **Is the project kosher?*** This involves more than simply determining whether the process for selecting a private developer is sufficiently objective to prevent the project from being handed to some politician's favorite nephew. There have been too many instances of state governments' selling an existing highway to a private firm for ready cash to plug a hole in the current year's budget, then leasing the road back for annual payments that burden future budgets. Privatization projects must produce meaningful benefits for all concerned—including the general public.

*2. **Does the project have widespread support?*** The surest way for a toll-road project to fail is for the general public to oppose it—because of anti-toll sentiment, a perception that the road is in the wrong place or not truly needed, or concern that it will have negative environmental consequences. It is too easy for community activist groups to stir up political opposition or stall the project in the courts.

A second guarantee of failure is for the public-agency sponsor to be anything less than fully behind the project. Officials who are unenthusiastic about a project have many ways to slow its progress until it lies dead in the water and the private developer decides to move on to better opportunities.

3. ***Is the project financially sound?*** A private toll road must be self-supporting. This means generating enough revenue from tolls (and possibly other sources) to cover all its annual costs and provide a fair rate of return on the private equity capital invested to build it. If the road is owned by its private developer, standard accounting rules require that the annual cost of depreciation be charged against revenues. This is a way of recognizing the diminishing asset value of a capital facility as it gradually "wears out."

To avoid having to set toll rates high enough to cover depreciation, legal title to a private toll road may actually be held by a public agency. For no sensible reason, public agencies in the United States are allowed to ignore depreciation as an annual cost, which eliminates the need for enough revenue to cover it. But at the end of the road's useful life, new debt must be issued to reconstruct it. In effect, depreciation is simply capitalized rather than being paid for by today's motorists.

4. ***Can the project produce significant benefits to society?*** The underlying rationale for any transportation facility is that it supports and generates economic activity by making possible more trips in less time. A higher level of economic activity today means a more prosperous society tomorrow. But tolls can discourage trip making on a road that charges them, especially when toll-free roads are available as reasonable alternatives (one of the problems experienced by Virginia's Dulles Greenway toll road).

Does this mean that we're kidding ourselves by imagining toll roads can provide a solution to our chronic pattern of under-investing in transportation capacity?

Not necessarily. California's State Route 91 toll road is demonstrating that many motorists will pay extra for a faster trip. The trick is to find a toll structure that produces enough revenue without discouraging too many trips. Sometimes this can require artful ways of hiding or ignoring certain of the road's annual costs so that toll rates can be lower. Capitalizing depreciation is one device for doing this.

Some people may be outraged by such blatant examples of "fiscal imprudence." But the important thing is whether the overall result is, on balance, beneficial to society. If we can pay for increased transportation capacity in ways that don't discourage its use, the result will be more trips and therefore more economic activity. Which is certainly beneficial to society, both today and tomorrow. And the size of tomorrow's benefits determines whether or not it is prudent to capitalize some of today's costs.

COSTS OF 5 PUBLIC-PRIVATE TOLL PROJECTS

FOOTHILL/EASTERN TOLL ROAD
28.6 miles of express tollways in Orange County, Calif. $1.5 BILLION.

SAN JOAQUIN HILLS TOLL ROAD
15-mile road in Orange County, Calif. $1.4 BILLION.

TORONTO ROUTE 407
Expressway/bypass for Toronto. $1 BILLION.

FARGO TOLL BRIDGE
Two-lane toll bridge between Fargo, North Dakota, and Moorhead, Minn. $1.6 MILLION.

TACOMA NARROWS TOLL BRIDGE
Bridge supplementing existing span in Tacoma, Wash. $800 MILLION.

PUBLIC-PRIVATE PARTNERSHIPS

Alternatives

Despite all the theoretical treatises that have appeared in learned journals over the years, transportation has to live in the real world. And its most real-world concern these days is how to pay for it. Which, in simple terms, comes down to a choice between taxes or user charges. In most cases, the choice we make depends on financial realities and on the cultural orientation of our societies.

The lone exception in the field of transportation is roadways. Until fairly recently, there has been no practical way for our societies to provide roadway services through the price system, which is the standard method we use for such other transportation modes as railways, airlines, canals and urban transit systems. So we have had no real choice except to depend on the tax system.

Not that the price system is automatically ideal in all cases and may not require some judicious help from the tax system here and there. But at least it provides a way to charge travelers for the amount of transportation services they consume. This has the benefit of requiring citizens to make personal choices about how much transportation makes sense for them to consume versus all the other goods and services they can buy, and the assumption is that such choices tend to be at least somewhat rational. The price system also provides an incentive for the providers of transportation to create the impression that their services represent "good value for the money" so they can attract more customers and maximize their revenues.

But until recently, this wasn't an option for roadway transportation. Apart from a few exceptions, there was no simple way to directly charge motorists for access to roadway lanes in the same way that we charge moviegoers for access to seats in movie theaters. And this led to the instinctive perception that "roadways were free," In other words, roadways weren't subject to the usual economic laws that require us to be at least somewhat rational in making choices about how much roadway travel to consume.

In most cases, the best we could do was to impose taxes on motor-vehicle fuels and assume the revenue collected would be a reasonable proxy for each motorist's roadway use. But because this form of proxy pricing divorces the act of paying for roadway use from the act of consuming it, motorists have never perceived it as a true price. And as every intelligent retailer knows, if you can divorce the act of paying from the act of buying (as with credit cards), you will always sell more. In addition, the use of fuel taxes as proxy prices tells us nothing about which roadway links are being used and when. All of this suggests our societies could be consuming more roadway services (relative to other transportation services) than makes economic sense and that we are flying blind in deciding how much to invest in each transportation mode.

As we know, the arrival of electronic toll-collection technology promises to change all this. We can now charge drivers "admission" to roadway lanes. We can charge them according to how many kilometers of roadway use they consume and what size vehicles they drive. We can even charge them based when they make their trips—with higher rates per kilometer during periods of the day when travel demand is high and lower rates during periods when travel demand is low.

The implications of these changes in how our societies can now pay for roadway travel services are far more profound than many of us realize. We're going to explore some of these implications to see what they can teach us. But first let us consider the context of what we're talking about.

PRIVATE GOODS VS. PUBLIC GOODS

Economists have filled many pages with discussions of the differences between what they have chosen to call PRIVATE GOODS and PUBLIC GOODS. All too often, the distinction seems to come down to a question of ownership. If a particular good is owned by society as a whole, it is a public good. But if it is owned by one or more individuals, it is a private good.

This distinction assumes special importance for many economic libertarians because of their concern over the rights of ownership, which they regard as one of the most critical aspects of free-market capitalism and its ability to function efficiently. However, it is of little help to us in our considerations of roadway transportation, where access is more important than ownership. Here is an example of why.

The vast New York metropolitan region is located on the western Atlantic coastline, so it contains many hundred kilometers of ocean beaches. In legal terms, each of these beaches is owned by the individual city or town within whose jurisdiction it lies. So most economic libertarians would automatically regard each of these beaches as a public good, without reference to the issue of access.

In New York City itself, access to Coney Island, Rockaway and all the other famous beaches is available to everyone in the world without charge—even to visitors from the U.K, France, Germany, Belgium and other members of the European Union savvy enough to make use of them in the summer. And the cost of operating and maintaining these beaches is funded out of New York City's general tax revenues.

This is not the pattern outside New York City. Instead, most individual towns charge fees for access to their beaches. Town residents pay an annual fee. Visiting non-residents pay a much higher (on an annual basis) daily fee. Ostensibly, this fee revenue is simply designed to cover the costs of beach operation and maintenance. But there are dark rumblings among social critics that some towns use their high daily non-resident fees to discourage beach use by lower income (and usually dark-skinned) people from other parts of the metropolitan region. While this may be illegal, it is very difficult to prove and is generally consistent with what social conservatives in the United States instinctively regard as the "American way."

In any case, this pattern of charging for access to beaches provides a down-to-earth definition of the most meaningful distinction between public and private goods without getting mired in the legal intricacies of ownership. A public good is freely available to everyone without charge and its costs are funded with taxes, while a private good restricts availability to those who are willing to pay for access.

Even such economic libertarians as Milton Friedman, Friedrich Hayek and Israel Kirzner recognize that most contemporary societies require at least some public goods in order to function effectively, even in a purely capitalist environment. The examples they usually give tend to be limited to law enforcement, especially those aspects of law enforcement involving business contracts. But to the man in the street, law enforcement means police services.

No one has ever been able to come up with a satisfactory model for delivering critical services like police and fire protection except as public goods. There is general agreement that they must be provided by government. Access to them must be free to all citizens without regard to their income or where they live. And their costs must be covered by general tax revenues.

This usually means that police officers and fire fighters are government employees, and their services are operated as government departments. The same pattern prevails for other public services provided by government, including roadways.

Interestingly, the costs of having police and fire services provided by government departments rather than private firms are often surprisingly modest because government is able to take advantage of certain sociological factors that are not available to private firms. Among working-class individuals in cities like New York and Boston, for example, being a police officer or fire fighter is regarded as a sacred family tradition to serve the public. In return for laying their lives on the line every day, all these public servants (many of whom develop astonishingly effective street smarts) ask is a decent living and a guaranteed city pension. There is no way that any private-sector police service provider could attract similarly qualified and dedicated employees except by offering salary and benefit packages comparable to the seven- or eight-figure dollar amounts typical of those paid by major professional sports teams.

But some societies have been experimenting with having government enter into contracts with private firms to provide roadway services, in order to benefit from the assumed greater operating efficiencies of the private sector. These experiments have covered a wide range of possibilities.

In their simplest form, government may contract with private firms to take over responsibility for a sub-function like roadway maintenance and be paid a fixed annual fee. At the other end of the spectrum, government may allocate to a private firm the entire responsibility for a particular roadway, including building and owning it. In return, the firm often receives annual payments from government that aren't fixed but vary according to such factors as traffic volume, which are regarded as "objective measures" of how effectively the firm is doing its job.

But no matter how simple or complex such outsourcing arrangements may be, the income of the firms involved in these arrangements comes entirely from the contract payments they receive from government. Therefore, even roadways that are privately owned and operated continue to have their costs funded by general tax revenues. So motorists aren't directly charged for using them and they remain public goods.

Many societies have taken this concept of outsourcing one step further. They authorize private firms to provide important services and charge users for access to them. Such firms are expected to recover all their costs (including a fair return on their assets) from these fees and receive no payments from government. This is a critical difference from the outsourcing model we just explored. Since users must pay for access to these services, this model converts what could be a public good into a private good. It is commonly followed for utility services—including electric power, water supply and telephone communications.

At one time, the common view was that the need to assure sufficient service capacity and reliability was more important than any benefits that might accrue from having private utility firms compete with each other in the marketplace. So government would franchise individual private firms to provide utility services to specific markets on a monopolistic basis. In return, the firms had to agree to have their operations regulated by government, which would set fee schedules that were designed to assure the firms of a "fair" return on their invested capital. This essentially converted these firms into little more than quasi-government enterprises—and they tended to be regarded as such by the general public.

But the growing intellectual popularity of the "deregulation" concept began to undercut this common view in societies like the United States during the 1980s and 90s. The feeling was that the benefits of more wide-open competition would result in greater efficiencies and lower prices to users, especially in traditionally regulated industries like utilities, while the "self-regulating nature" of free-market capitalism would assure adequate capacity and reliability. So it was that a passion for deregulation swept the United States.

Unfortunately, free-market capitalism (like organized religion) has a long history of bringing out the worst in people. The siren call of higher profits led top managers in corporations like Enron to rig markets in order to create artificial shortages that enabled them to boost prices, to invent a host of shadow corporations to absorb certain operating costs so that their income statements would look more attractive, and to engage in other forms of chicanery that represent capitalism at its worst.

One result of such greed-driven mismanagement was a major electric power failure in 2003 that disrupted the lives or more than 50 million people in the Northeast and Midwest sections of the United States.

Another is the fact that the formerly regulated domestic airline industry has become a financial basket case under deregulation—to the point where it has lost more money than it ever made and the continued viability of this critical U.S. transportation network has been called into question. There is growing awareness that the true costs of deregulation are far higher than most Americans anticipated.

THE HONG KONG MODEL

An interesting variation of the classic regulated utility model can be found in Hong Kong. Its government created and capitalized three commercial corporations to operate its new international airport, its high-volume metro system and its suburban railway system. Each of these corporations is expected to function as a profit-seeking enterprise in the narrow accounting sense, but with the government holding the policy reins to make sure that they generate "social profits" as well.

This model has been notably successful in a city that prizes its reputation as the world's freest capitalist economy. All three corporations are nicely profitable and the transportation services they provide are generally regarded as among the best in the world. And in the best venture-capitalist tradition, Hong Kong's government has begun a process of recovering the capital it invested to get these corporations up and running by selling minority ownership shares in them to private investors through the local stock market.

Three Important Lessons

So far, we've chewed our way through a mouthful of world experiences with delivering essential services to society as either public goods or private goods. While the record is a mixed one, at least three lessons emerge.

One is that delivering most of these services as private goods through some kind of price system has much to recommend it. Not the least of its benefits is the discipline it imposes on society to make intelligent use of essential services—neither engaging in wasteful consumption of them because they appear to be free, nor starving itself and its economic growth of an adequate supply in a misguided attempt to "save money." When individual members of the public must pay for services according to how much they consume, each person is more likely than not to make rational trade-off decisions about how much consumption makes economic sense to them personally at any given time. In turn, the collective impact of all these decisions, expressed through a fairly wide open marketplace, creates useful economic signals that guide society in determining how much capital to invest in the facilities needed to provide any given level of service capacity.

The second lesson concerns capitalism's notorious "unruliness" in the absence of external constraints, which can do serious damage to society. An extreme example occurred during the 1920s in the United States when the sale of alcoholic beverages was prohibited by law. This led to a large and thirsty underground demand by American society for such beverages, which was met by criminal enterprises operating under what might be called "the Al Capone version" of Chicago capitalism.

These enterprises maximized their sell-side profits by dictating to the many illegal drinking establishments in the neighborhoods they controlled how much product each was expected to purchase each week and the prices they were expected to pay. Those who were unwilling to cooperate were usually bombed out of existence.

At the same time, these distribution enterprises minimized their buy-side costs by dictating to local underground breweries and distilleries how much product they were expected to supply and at what cost—or else their loaded trucks were simply highjacked. And to assure minimum interference by government in these illegal activities, the distribution enterprises bribed local police and the political establishment to look the other way. Needless to say, the resulting profits were enormous and paved the way for the modern organized crime cartel in the United States that is popularly known as "the Mafia."

The third lesson we should consider involves the issue of economic justice. This relates to the "willing buyer and willing seller" concept that many libertarians glibly assume as a given in their discussions of the marketplace.

But it is scarcely a given to the homeless man standing in the rain with his shivering family seeking shelter from an apartment-building landlord who demands an unaffordable rent on a take-it-or-leave-it basis. In the case of many such non-discretionary needs, there is no such thing as a willing buyer, and the ability of the marketplace to serve society is gravely crippled.

Hong Kong and Economic Justice

One of the most interesting ways of dealing with this problem can be found in Hong Kong. During the decades following the end of World War II, Hong Kong's population experienced rapid growth because of a massive influx of immigrants from the Chinese mainland. These repeated waves of new residents were mostly from rural backgrounds and had low incomes. Their burgeoning numbers created a huge demand for low-cost housing that Hong Kong's private real-estate industry couldn't begin to meet. The inevitable result was the growth of vast, informal, Third World shantytowns that spread like virulent skin rashes up and down the previously vacant hills just beyond Hong Kong's built-up areas.

Some thirty years ago, Hong Kong's government decided to address this problem by effectively making itself responsible for meeting the society's non-discretionary housing needs through a major expansion of its housing programs. This decision led to construction of public-housing estates on a scale beyond anything contemplated by urban societies in Western Europe and the United States. It also liberated the private real-estate industry from the profit-constraining social-welfare baggage that is unavoidable when the private sector tries to satisfy all housing needs. Instead, it was left free to maximize its profits by concentrating on serving the purely discretionary housing demands of Hong Kong's more upscale residents through a largely unregulated free market.

Today, Hong Kong's population has stabilized at just below seven million. Approximately half of these residents (not just the low-income minority) live in some form of government-built housing. In most cases, their monthly housing costs are pegged at a fixed percentage of their family incomes—averaging an extremely modest nine percent. So these residents have a larger share of their incomes available to support the local retail industry than do their Western counterparts. Not to mention a dampened incentive to push for higher wages from their employers.

Beyond the obvious economic justice implications of Hong Kong's approach to housing lie some profound insights for capitalism generally. The free market works best when it can concentrate on serving society's discretionary demands for goods and services. This is the real meaning behind the oft-cited libertarian mantra of "willing buyers and willing sellers."

But no society is free of non-discretionary needs, and these needs must be met in an effective manner if capitalism is to enjoy sufficient latitude to work its wonders. We can choose to meet them through the mechanism of public goods funded by taxes and administered by some sort of central planning process. Or we can meet them by developing marketplace solutions.

Central Planning vs. the Marketplace

The central planning concept has fallen on intellectual hard times in recent years, and its critics like to cite the economic failures of Leninist Eastern European nations as proof that it doesn't work in practice.

But we should remember that the U.K and US—two of the world's most forthright capitalist nations—unhesitatingly chose central planning as their primary economic management tool to marshal the resources needed to win World War II. And both nations rightfully consider this among their proudest triumphs.

We should also keep in mind that throughout the world's capitalist nations, most economic transactions take place within the confines of large, vertically-integrated corporations—where they are subject to top-down central planning in a command-and-control environment that would do credit to the Leninist ideal. These two historic realities should caution us against automatically dismissing the option of meeting our non-discretionary needs with public goods provided through a central planning process as "ideologically unsound."

How To Provide Roadway Services

All this provides useful intellectual background for the choices we must make about how best to meet the needs of our societies for effective roadway services. Should we continue to provide these services as public goods that can be used by all drivers without charge and whose costs are funded through the tax system? Or should we take advantage of the new freedom that electronic toll collection now offers us to provide these services through the marketplace as private goods whose costs are funded by user charges? Is it possible to combine the best of both options?

The marketplace approach may offer special benefits in the major metropolitan regions that have the largest aggregate travel demands and generate most of our societies' gross domestic product. Under this approach, we would implement user charges on the limited access highways serving these regions. But if user-charge rates per kilometer are set at economically sensible levels, they could generate enough revenue to fully support a metropolitan region's entire roadway system—including its local streets and avenues as well as its limited-access highways. Doing so would convert these essential transportation systems into self-supporting enterprises.

This would enable us to remove the cost of providing roadway services from all funding dependence on government budgets, which are increasingly pressured by the growing needs of aging populations and other costs that can only be funded through the tax system. It would also assure metropolitan roadway systems of the revenues they need to provide better service to motorists and meet capital needs without cutting corners, thereby improving their effectiveness in supporting economic growth.

Our governments could franchise private firms to operate these self-supporting roadway enterprises as regulated utilities while retaining ownership of the roadways themselves. Or they could transfer both operation and ownership to the private sector by selling off the roadways.

But a more effective option may be to adopt the Hong Kong model—having government create new, government-owned commercial corporations to own and operate metropolitan roadway systems. Doing this would make it possible to attract private investors to provide equity capital in exchange for a minority share of ownership. Some of these investors would be attracted simply by the promise of investment income. But other investors could be electric power and telephone utilities, retail-banking corporations and other large companies whose revenues are keyed to economic growth in the metropolitan regions where they do business—which gives them a special interest in making possible better roadway transportation.

In any case, the presence of such "patient equity" in the capital structure of these roadway enterprises would reduce their need for debt capital in funding the costs of capacity expansion and renewal of existing facilities.

Economic Justice

Charging motorists for using limited-access highways raises certain economic justice issues that must be addressed. The most obvious issue is that distance-based highway charges can impose a difficult burden on the working poor, who may not be able to use public transportation to travel to and from their jobs and have no option other than driving cars. Daily trip charges may be too small a proportion of family incomes to matter much among middle-class and upper-class workers, but this proportion can be a significant burden for the working poor.

A common solution to this problem is the old Leninist approach to pricing consumer goods and services—keep trip charges low enough to be affordable to everyone. But this solution doesn't address the issue of the relative proportion of family income that these costs represent. It can also result in under-pricing highway access to society as a whole, which can defeat the purposes of charging for highway use in the first place.

At the other end of the spectrum is Hong Kong's policy of charging its public-housing residents a monthly rent that is pegged to some fixed percentage of their family incomes. This assures that all residents have roughly the same housing-cost burden relative to family incomes even while the cash cost of this burden may vary widely among residents of identical housing units. Like the progressive income tax (which has become regrettably unfashionable in many of our societies), how much you pay depends on your ability to pay—which reflects the size of your income and therefore tends to be a good measure of how much you are benefiting from society in financial terms.

In the case of highway charges, the simplest solution may be to take advantage of the flexibility that is built into electronic toll-collection technology. When a motorist makes a trip on a particular highway, this technology measures the distance he travels and then charges his account a fee based on a particular rate per kilometer. Since this process is unique to each motorist, the rate he is charged can also be unique—making it possible to provide discounts to motorists who can establish that their income falls below a certain threshold level. There can even be a sliding scale of discounts reflecting differences in income.

The ability to charge motorists different rates per kilometer makes it a simple matter to incorporate a host of other economic justice factors in determining these rates. For example, a heavy truck obviously causes more pavement wear per kilometer than a lightweight automobile. So it is only logical to charge the truck a higher rate per kilometer than the automobile. And if the automobile happens to be one of the new hybrid vehicles that generates far less air pollution per kilometer than standard automobiles, we may want to charge it a special low rate.

We can even use this differential rate concept to manage travel demand far more effectively than has been possible in the past. As all of us are aware, a highway that is congested with too many vehicles experiences a reduction in the average speed of the traffic flow, and this reduces the number of vehicles per hour that the highway can process. But we can use differential pricing to shift excess demand to less crowded roadways and to less crowded periods of the day. In effect, we can achieve any average traffic flow speed we wish on a highway link by adjusting the rate per kilometer on a real-time basis in response to demand.

Doing so makes it possible for us to guarantee to our motorist customers a relatively high average traffic speed on our highways. This means we are providing them with better service than is now possible because it reduces their trip times. Also, reduced trip times means vehicles spend less time generating air pollution, which is an increasingly important social benefit. All of this is possible when we exploit the potential of electronic toll collection technology in imaginative ways.

INTERMODALISM

An increasingly popular topic in transportation journals these days is the concept of intermodalism. There is growing awareness that the challenges of 21st-century globalism require us to operate our roadway, rail and water transportation networks so that they complement each other as components of an integrated surface transportation system that improves travel efficiency. For too long, we have operated these networks as if they existed in separate universes.

An obvious example is the growing use of large trucks for goods movement on the congested highways of our major metropolitan regions and in the travel corridors that connect them. This is a particular problem in the United States, where the privately owned freight railroads have been shrinking their track capacity in order to reduce maintenance costs. The result is that increasing amounts of freight that should move by rail ends up in trucks because the freight railroads lack sufficient track capacity. This reduces transportation efficiency and imposes unnecessary burdens on society.

Clearly, a more rational allocation of demand among modes that are part of a true intermodal surface transportation system would pay economic dividends. But where do we get the money to fund such integration?

As it happens, our roadways are the only surface transportation mode whose services have never been provided as private goods through the marketplace. If we change this by exploiting the ability of electronic toll-collection technology to price access to them in an economically rational manner, some of the revenue generated can be used to help fund progress towards intermodalism. This could include paying for more track capacity on freight railways so they can accommodate more goods volume, which can make a lot more sense than blindly constructing more highway lanes to handle an increasing flood of trucks—especially in dense metropolitan regions where rising social concerns about protecting developed neighborhoods can impose stringent limits on taking more land for highways. The same considerations apply to developing more capacity on traditionally under-financed public transportation systems in metropolitan regions.

Conclusions

The nature of our societies means that technology usually advances more rapidly than our ability to make intelligent use of it. So the most important challenge we face in enabling our roadways to serve us more effectively is to broaden our horizons when it comes to managing the potential inherent in electronic toll-collection technology.

Obviously, we can sidestep this challenge by continuing to provide roadway services as a public good, funded through the tax system. But this will only work if tax revenues are robust enough to assure roadway systems that are able to accommodate rising transportation demand. In the United States, and possibly other Western societies as well, this no longer appears to be feasible except by increasing tax rates or shifting existing tax revenue to roadways from other government services—both of which open some awkward cans of worms.

Or we could try to "manage demand" in ways that reduce growth. But since transportation demand is a natural consequence of economic activity, this means turning our backs on the kind of economic growth needed to assure better living standards for our societies.

A more sensible choice may be to charge motorists for roadway use, thereby creating a new source of badly needed funds for transportation.

Public–Private Partnerships: Brave New World

Americans take millions of trips every day—to work, for pleasure, and from errand to errand. The nation's system of roads is one of the world's most impressive public investments and has contributed greatly to our economic success. At the same time, Americans are proud of our commitment to a market system that has afforded us a standard of living envied by the entire world. So why maintain the belief that only the public sector can provide for our transportation needs?

The source of this belief is certain assumptions made about who is responsible for delivering and maintaining the road system and who should pay for it. Roads are assumed to be a public good. A public good is a product or service that everyone needs and from which everyone benefits, but that private firms to not have sufficient incentive to produce; thus the market will supply either an insufficient quantity of the good or none at all. This stems from the fact that no one can be excluded from enjoying a public good, and that one person's use of such a good does not affect anyone else's use of it. The most common example of public good is national defense. If one person benefits from having an air force or an army, everyone does.

The truth is that much of the nation's transportation system should not be treated as a true public good. When transportation planners tell us that ridership on a particular strip of roadway is expected to increase dramatically, we know that the cost to maintain the road will increase as well. Adding one extra-heavy truck will put a strain on the system and decrease the life of our infrastructure investment. Moreover, the nation's

---

Adapted from Joseph M. Giglio and William D. Ankner, "Public-Private Partnerships: Brave New World," TR News *(Washington, DC: Transportation Research Board) 198 (September-October 1998): 28–33.*

transportation system is commercially viable. Many roads and bridges can be priced to benefit their intended users, allowing private firms to assess the commercial viability of undertaking certain infrastructure projects. When highways, for example, are viewed as a public good, there is a tendency to treat them as a public charity, placing highway professionals in the position of being landlords instead of managers. Transportation is neither a true public good nor a purely private one.

Indeed, responsibility for transportation investment lies with a mix of public, public-private, quasi-public and private entities. Responsibilities for the various transportation modes have shifted over time. Early roads in this country were private and were often toll roads. Bridges were private and tolled; some such bridges still exist. The railroads were private for both freight and passengers; today rail freight is private, and passenger service is quasi-public. Subways in New York were initially private. Airports also started out private; some remain so, but most have shifted to local governments or quasi-public entities. To deal with these myriad levels of responsibility, various schemes have been devised for the relevant entities to finance their investments—from reliance on market forces to the imposition of user fees and tolls. Some have worked, while others have not.

GROWING INTEREST IN PRIVATE INVOLVEMENT

The biggest driving force behind the new interest in private involvement in transportation has been the changing nature of the federal-state partnership.

America's transportation network is the largest and most complex system in the world. Construction and maintenance of this system requires massive amounts of public spending each year. Yet in 1993 total spending by all levels of government fell $16.7 billion short of the investment needed just to maintain the current system and $39.5 billion short of the amount required to improve conditions. The U.S. Department of Transportation estimates that approximately $249 billion will be needed during the next five years simply to maintain the existing condition of the nation's highways. The Transportation Equity Act for the 21st Century (TEA-21), recently passed by Congress, does not meet the level of investment necessary to maintain the system or handle growing capacity demands. As we approach the 21st century, our ability to compete globally depends on our willingness to make critical domestic transportation investments.

Private-sector involvement in transportation is nothing new. Transportation agencies have used private firms extensively for construction, maintenance, design and engineering services. Private firms have assisted governments in preparing designs for public works projects and providing utility services as regulated monopolies. But prior to recent federal legislation and several local initiatives, only a limited number of arrangements existed with private firms concerning ownership and operation of highway facilities.

As budget realities are brought to bear, all levels of government are looking for new ways to finance increased infrastructure investment. The gap between revenue and needs grows wider every year. Filling that gap typically falls on state and local governments, where fiscal pressures appear likely to continue for many years, independent of current budget surpluses. This trend shows no signs of abating, as demands on the road system increase while the need to limit government spending grows. Governments at all levels are seeking service-delivery approaches that can contain costs and improve the quality of services provided. Unfortunately, the gas tax is limited in its ability to finance the gaps, and the returns will diminish as we achieve other social goals such as environmentally and energy-efficient automobiles, trucks and buses. Thus we need to look at new opportunities.

Controlling costs and providing better services should be the public policy goals of any government agency, regardless of budgetary constraints. For many reasons, however, public agencies are not always able to deliver on these goals. Greater involvement of the private sector is one way to contain costs, improve service and free up some room in government budgets for true public goods.

Public–Private Partnerships

Public-private partnerships offer the hope of enabling the transportation industry to increase the number and amount of transportation investments. Initial efforts, however, have not been promising. The public sector's approach is to use the private sector as a contractor, not as a partner. Contracting out work does not include the private sector in the investment, nor does it involve private equity, and thus the private sector has limited incentive to generate the best return for the project. The public sector monitors, inspects and oversees the private sector to ensure that the project is accomplished correctly. The relationship between the participants is too often one of distrust.

An opportunity within public-private partnerships is the development of a real "equity partnership" approach. This approach cannot be used for all or even most projects, but it can be applied for some and often for larger projects. Under this approach, the public and private sectors share in both the risks and profits of the project over the long haul. Each brings its strengths to bear. For example, the public sector has the best resources for performing the up-front and high-risk work of project development, environmental assessment, community outreach and condemnation. The private sector's contribution is efficiency and quality. Under the equity partnership approach, ensuring efficiency and quality is in the private sector's interest; therefore, the public sector need not worry about overhead rates, quality control or inferior materials. Thus public-sector overhead costs can be lowered, thereby reducing the overall cost of the project. The public sector also shares in the project's profitability. This is especially relevant when the public sector takes the up-front risks.

Risk and Return

A key factor in any investment, whether public or private, is the assessment of risk and return. Simplistically, the return for the private sector is seen as monetary—either profits or increased market share/penetration that translates into profits. For the public sector historically, the return is not money, but various social or public goals such as mobility access and economic development. Risk is also bifurcated. The private sector views risk as a threat that must be overcome by a larger opportunity for profits; for the public sector, risk is part of doing business to meet public goals.

These contrasting views have led to the idea that it is wrong for the public sector to earn a monetary return on investments and for the private sector to make social goods equal to profits. This rigidity has led to a number of absurdities, particularly from the public side. For example, the building of the Interstate system has increased mobility and economic development in the United States. Throughout the nation, new businesses and investments cluster around Interstate interchanges. However, the public sector is not allowed to own the land around the interchange and share in the interchange's development. The public sector takes the risk and receives none of the economic development benefits associated with its investments. The result is that money is made, usage at the interchange increases, and the public sector has to maintain the interchange and often make improvements to meet the increased demand.

These lost opportunities go beyond highways and extend to other modes as well. For example, transit is often the biggest victim of this public rigidity regarding profit. The transit infrastructure of New York City is essential for the city's economic vitality. However, transit is viewed as a financial sinkhole because it shares in none of the economic development profits. For example, the subway and PATH service into the World Trade Center complex allow that complex to exist profitably. However, none of the mall revenues or the office rentals in the complex are credited to transit. Consequently, PATH is a deficit operation of the Port Authority of New York and New Jersey, and the World Trade Center is viewed as a "cash cow."

Transportation investments, whether in highways or other modes, create value. The public sector should be able to capture all or some of that value to help finance and maintain the transportation investment and the overall transportation system.

Public or Private Sector?

We have in fact a surprising amount of leeway in deciding whether roads should be built and operated by the public or the private sector. The decision can be made on a road-by-road basis and comes down to which sector can deliver the best overall benefit to the public for the best value. How can a profit-seeking, tax-paying private company deliver services at lower cost than public agencies and still make a profit? At least five factors can enable private firms to deliver services less expensively or provide greater net benefit than public entities.

Greater operating efficiency. Greater operating efficiency means getting more return on one's investment. This results partly from good management, but managers must be free to exploit every opportunity to achieve greater efficiency—a luxury public agencies do not always enjoy.

Public agencies must make every effort to avoid the appearance of fraud or favoritism in their purchase of supplies and services. The result can be elaborate bidding procedures for purchase contracts, lengthy audits before bills are paid and large amounts of time-consuming paperwork. Public-agency managers rarely have the option of simply picking up the phone and choosing a supplier based on timely delivery, quality products and attractive prices. They usually must restrict themselves to a small group of suppliers that have long-standing government contracts and can wait months to be paid. All of this means higher costs for every purchase contract.

The bottom-line focus of most private companies forces them to operate in a world in which elaborate procedures do not interfere with practical results. Private-sector managers can exploit the benefits of just-in-time inventory management, economies of scale in purchase orders and meaningful measures of supplier performance to make their operations more efficient.

Greater efficiency in service delivery also means taking risks, including being aggressive in exploiting the potential of new technology to streamline production and reduce costs. Private managers are paid to accept and manage risk. Their public-sector peers, however, generally have no such incentives. Preferring not to "risk the public money" until technology has gained industry acceptance, for example, public agencies are usually last to adopt modern technology and end up with outdated systems.

Absence of conflicting goals. In theory, a public agency should have only one goal—to deliver effective services with maximum efficiency. In practice, however, this goal can often be complicated by others. These other goals may include various social benefits, such as providing more entry-level jobs for young people, higher community standards for wages and benefits and improved opportunities for women and minorities. While important in their own right, these goals have the potential to conflict with an agency's goal of service delivery, and can result in higher costs for both public services and the social benefits they are intended to generate.

Politics also plays a big part in the administration of any public agency. Public-sector managers are frequently subjected to political interference, from which a private firm is generally free.

The marketplace can often provide a practical solution to these potentially conflicting goals. This can occur when public agencies and private firms are allowed to compete on price for contracts to deliver public services. In a surprising number of cases, such open competition has enabled public agencies to become as cost-efficient as the best private firms. In other cases, private firms have found it in their best interests to shave their profit targets in order to win service contracts. Either way, willingness to let the marketplace help choose a service provider leads to the kind of environment that results in better public service at less cost.

Easier access to low-cost capital. Most public services, including roads, require investments in capital facilities. First the roads must be built. Then they must be maintained and eventually rebuilt or expanded, depending on travel demand, and this requires capital investments over the life of the project. One factor enabling private participation in a way that provides the best overall package is the private sector's access to capital.

Public agencies subject to debt restrictions. State and local governments are often subject to restrictions on the amount of debt they can issue and for what purpose. In some cases, debt issues must be approved in advance by voters. Given the public's aversion to debt, such referendums are often difficult to pass. In still other cases, elaborate formulas may be built into state constitutions, linking the maximum amount of debt a county or city may have outstanding to a measure that approximates the locality's ability to support that debt, such as the value of taxable real estate. As a result, state and local governments may not be able to raise enough capital to build the roads and other public facilities required by to day's economy.

While some states have tried to work around these debt-issue restrictions, the overall effect of such efforts tends to be marginal. For example, a state turnpike authority may be able to issue an amount of debt equal to its toll revenue. Using an amendment to its charter (or simply a less rigid interpretation of its charter), the authority uses these new funds, supported by tolls, to build or expand toll-free roads that are expected to increase traffic on the toll road. There are several such means that can be used to work around debt restrictions, but taxpayers' general aversion to obligations will always limit the impact of such measures.

Over the years, the federal government has attempted to fill the capital gap through its various transportation programs. The present federal transportation legislation provides states with more flexibility in response to budget constraints. Current law allows federal dollars to fund up to 80 percent of the cost of an "eligible transportation project." While this sounds rather open-ended, the annual federal appropriations process, coupled with deficit-reduction guidelines, severely limits the amount of federal aid available.

The market for taxable debt dwarfs its tax-exempt counterpart. Because the private debt market is so much larger, private firms can bring a wealth of investment dollars into transportation finance. Projects that formerly had to compete for a limited amount of public funds thereby have access to a much larger pool of capital that comes with the involvement of the private firm.

Traditional finance theory argues that all debt should be self-financing. The capital facilities it creates should generate a new income stream to cover interest and principal payments by a comfortable margin. It may be fiscally prudent, however, to issue increasing amounts of debt in order to exploit whatever new economic opportunities are able to satisfy the self-financing mandate. The overall net benefit to a state transportation agency may be greater than if the agency uses its scarce funds for projects that cannot be self-sustaining. At the same time, a role is carved out for private capital in those projects that can be self-sustaining.

Tax subsidies to private firms. There is no shortage of capital-seeking investment opportunities. Large corporations, unencumbered by the fund-raising restrictions that limit government, have access to capital as long as they can demonstrate their ability to generate new revenue. Many corporations have relied on this open access to capital in order to finance the development of new products they expect to sell at a profit. What if these products happened to be roads or other public goods traditionally regarded as the exclusive purview of government? Allowing these firms to finance the development of public works would go a long way toward meeting the financial needs for the infrastructure upon which the nation's future prosperity depends.

While access to capital is an important consideration, keeping the cost of capital as low as possible is fundamental in determining how best to finance infrastructure needs. A state or municipal government can raise capital funds for its public agencies by issuing bonds whose interest payments are not subject to federal (and in some instances state) income taxes. Thus these tax-free bonds should enjoy lower interest rates than the taxable bonds issued by corporations, and they generally do.

However, the federal government provides two subsidies to private companies that are not available to public agencies. Under the right circumstances, this can have a significant impact in reducing a private company's true cost of capital. When a private party issues the debt itself, one of these subsidies is the full deductibility of interest payments in determining the company's taxable income. For many profitable companies, this can result in out-of-pocket (or after-tax) interest costs that are competitive with tax-free debt. Since public agencies have no income tax liabilities, their normal interest costs are identical to their out-of-pocket costs.

The second federal allowance concerns a company's depreciation deduction—the prorated cost of a capital facility over the years of its useful life. Since all deductions for depreciation mean lower taxable profits, a company's actual profit (return on invested capital) is considerably higher than the profit it must report to the Internal Revenue Service.

These tax deductions for interest and depreciation flow directly into a firm's bottom line. In effect, they may reduce the company's cost of capital to levels that are comparable and competitive with tax-exempt debt issues of public agencies.

What Role Should the Public Sector Play?

While the private sector can provide real advantages by participating in transportation services, there will always be important functions more appropriately carried out by the public sector. In particular, the public sector is key to ensuring that:

- Service levels and standards are maintained.
- Social goals are achieved.
- The best financial package is used, whether that means private, public or a combination of both.

Maintain service levels and standards. Because substantial amounts of money are involved in many traditional transportation projects, there may be an incentive for the private sector to cut corners in order to lower costs and thus increase profits. Government must therefore continue to set guidelines and enforce standards in order to maintain quality and ensure the safety and welfare of the public.

To protect quality, performance and incentive requirements are usually written into agreements with private firms. Besides a formal contract requiring the private partner to maintain a specified level of service, the government must have a mechanism in place for adequately monitoring the performance of the private firm. In addition, the government can reduce the possibility of corruption (kickbacks, fraud) by establishing sound but streamlined procurement procedures or, in the case of divestiture of responsibilities, instituting necessary regulations and oversight.

The public sector's oversight responsibility is different if the private and public sectors are true long-term equity partners. This is because the private sector can sustain long-term profits only when quality control and efficiencies are built into the beginning of a project.

Private entities may be more likely than the public sector to reduce or interrupt operations as a result of financial problems or strikes. Service interruptions can be of particular concern if they affect the safety and welfare of the general public. Government must therefore consider which of its operations are most critical to the needs of the public and, if necessary, continue to provide these services on a public basis.

All levels of government, particularly federal, also have a role in ensuring the systemwide integration of the nation's transportation network. While states and localities should be free to pursue innovative means of transportation finance, each individual network must be consistent within the entire system. Imagine, for example, 50 or more different tolling mechanisms, one for each state, across the country. Public agencies must therefore play a critical role in establishing guidelines to promote consistency and efficiency within the system.

Achieve social goals. Because private firms are profit oriented, they will be tempted to avoid (or reduce) the level of service for those segments of the general public that are unable to pay for service or may be expensive to assist. All levels of government have a role and responsibility to aid the disadvantaged and ensure their needs are met.

Use the best financial packages. As noted earlier, there are opportunities for private firms to issue taxable debt that is competitive with government's tax-exempt equivalents. But they may not always exploit these opportunities. In many instances, firms do not want to issue the debt or cannot raise sufficient funds for the project. This situation may stem from the level of risk involved in the project or uncertainty surrounding traffic volume.

While privately issued (taxable) debt usually requires 20 to 25 percent equity from the participants, public financing enables full 100-percent project financing because the municipal market is accustomed to such financing structures. Private participants may seek out an existing public authority to act as the bonding agent, ask the DOT (or state) to issue tax-exempt general obligation or revenue bonds, or offer to set up a special-purpose nonprofit corporation for the project's duration. Depending on the project, these special-purpose entities may also issue tax-exempt, non-recourse revenue bonds. In this way, the public sector may play a key role in financing the project.

Overcoming Obstacles to Private Involvement

The ideas presented in this article represent a strategic shift from the traditional practices of transportation finance. No doubt these new approaches to highway financing will face opposition from defenders of the status quo in both the public and private sectors. Yet private-sector involvement can coexist along with more conventional financial arrangements, and a commitment to innovative methods can provide a transportation network capable of meeting the demands of the 21st century. First, however, several obstacles must be overcome.

Mind-set. Probably the single largest impediment to greater use of the private sector for the delivery of transportation services is a lack of public experience with these alternative approaches. Not until we move beyond the current paradigms concerning the limits of public and private provision of such services can private involvement in transportation really work.

As discussed earlier, the government and private companies must be seen as natural partners in providing the wide range of public services our society needs if it is to prosper in the future. Government has the obligation (and the authority) to set standards for service delivery that are focused on the welfare of society as a whole, rather than the narrowly defined economic welfare of any particular company. The equity partnership arrangement thus requires competent, flexible public-sector managers to create efficient ground rules and design a regulatory framework. The private sector has the ability to meet these standards efficiently and to attract the capital funds necessary for the modem infrastructure that is essential for effective service delivery.

DOTs view themselves as landlords, building and occasionally maintaining their transportation systems. This paradigm must shift so that DOTs take on the role of managers. Transportation is a business, and its financing is part of management.

Institutional barriers. Facilitating an alliance between the public and private sectors will take more than a few individuals championing a particular project. Governments will need to establish both an organizational and analytical framework for the process. Guidelines, organizational arrangements and knowledgeable staff must be put in place to ensure the success of a public-private paradigm.

Although much work has already begun, a great deal more needs to be done. Important issues such as the types of programs to implement, the degree of government financial support and the required regulations should be considered. Approaches also must be tailored to particular political, economic and labor environments. Several state governments, for instance, have established a government-wide commission to identify public-private opportunities within the state's agenda and establish policies for the state's own initiatives.

Legislative and political barriers. Several opportunities for public-private partnerships have been created by existing policy and legislation. Many states and localities, however, have yet to adopt the legislation that will enable them to include the private sector. This can take place only when such legislation has the support of committed political leadership. Establishing this crucial support early on will sustain momentum for these arrangements and make it possible to address any concerns as they arise.

PUBLIC-PRIVATE PARTNERSHIPS

Private Roads and Public Benefits

Is there any sensible reason why we Americans should continue to maintain a Chinese Wall between public and private goods and who is allowed to provide which? Or are we simply making life unnecessarily difficult for ourselves?

Who Owns What?

The driveway that leads to my garage was built by the private real-estate developer who sold me my house. Because I own it, academics like to call it a private good. I am responsible for clearing it of snow in the winter and otherwise keeping it operational. Whatever costs are involved in doing this I pay directly out of my own pocket.

The local street that leads to my driveway was built by my town's government. It is owned by the town, also serves my neighbors' driveways, and is therefore what academics call "a public good." The town is responsible for keeping it operational, which includes plowing it free of snow in the winter. These costs are covered by the town's revenues from real-estate taxes—which means that I pay for them indirectly.

This distinction between what people own individually and what they own collectively is the rule in most parts of the United States. It leads to the knee-jerk assumption that the roads and other public goods we own collectively have to be the direct responsibility of government, while responsibility for the driveways and other private goods we own as individuals should be ours alone.

Adapted from Joseph M. Giglio, "Private Roads and Public Benefits," The World & I *12, no. 9 (September 1997).*

But things are different in the new housing development on the other side of town where my friend Tom lives. The local street that provides access to his and his neighbors' driveways is owned by the private real-estate company that built the houses. The responsibility for keeping this public good operational rests entirely with the real-estate company, which bills Tom and his neighbors each month for this service. Because the town government doesn't have to spend tax dollars maintaining the street, Tom and his neighbors have lower real-estate taxes than I do.

When it snows in the winter, the real-estate company must naturally plow Tom's street. But it also plows each homeowner's driveway at the same time. The extra cost to the company of providing Tom and his neighbors with this attractive and ostensibly free service for their private driveways is insignificant because its snowplow is already on the scene to plow the public street.

I sometimes envy Tom, especially when I have to shovel out my driveway after a heavy snowfall. Of course, I could always pay someone to plow my driveway. But there's no way to coordinate this with the town's snowplowing schedule. So I might still end up having to shovel out my driveway's entrance after the town's snowplow has plowed it shut.

THE LOCAL HIGHWAY MESS

The need to find a better way of managing public roads is painfully apparent to me each morning when I drive to work. Most of my commute is on a highway that has been badly neglected by the county since it was built back in the 1960s. Its pavement is worn and rough. Its four lanes aren't really wide enough for the new generation of trailer trucks that travel them in increasing numbers. And there is a two-mile stretch lined on each side by strip malls and other retail establishments, most of which were built during the 1980s. The eight traffic lights along this section are an unending source of delay for commuters like me during rush hours.

The highway's many shortcomings are a favorite topic for critical stories in our local newspaper. Why can't the county maintain the pavement in better condition? What happened to the plan announced five years ago to widen the lanes? Isn't it possible to build a bypass around that two-mile commercial stretch so through traffic can avoid those eight traffic lights and all the cars pulling out of parking lots in front of the stores?

Our newspaper delights in putting these questions to county officials. And their answers are always the same. Not enough money in the budget.

Too many competing demands for county services. This year's proposed tax hike was vetoed by the county's elected board of supervisors. The state government always shortchanges the county in disbursing roadway funds from the statewide gasoline tax. But underlying every newspaper story that prints these standard excuses is the same implicit theme. Better management might make all the difference. New solutions are needed.

What would happen if the highway were turned over to the same private company that owns and operates the local streets in Tom's neighborhood? This question keeps haunting me as I struggle with stop-and-go traffic along the highway's commercial strip twice each day. Could such a company build the long-needed bypass and pay for it with a modest toll that commuters might welcome in exchange for a faster trip to work? Would the toll revenue be enough to fund other improvements to the highway as well?

The Privatization Controversy

There is a typically clumsy academic name for the way local streets are managed in Tom's neighborhood. It's called "privatization." It means having private firms become responsible for certain public goods that may traditionally have been the responsibility of government. The advantages? Better services and lower costs to the public, claim its advocates. The disadvantages? The biggest one may be the need for us to take a fresh look at that imposing Chinese Wall between the public and private sectors. Nontraditional thinking is always such a burden.

Privatization is really just a new name for an old concept. Most governments have always retained private companies to design and build public facilities like roads and schools. My town government "contracts out" to a private company the task of maintaining our local street-lights and traffic signals. In fact, this company has been providing the same services to most towns in my area for years, and no one considers this strange or radical.

The new wrinkle in private operation involves applying this concept to public services that have long been regarded as the exclusive preserve of government. That's why there was so much controversy in my town five years ago when the real-estate company submitted its application to build the housing development where Tom now lives. The town government wanted the additional real-estate-tax revenue that the new houses would generate. But its budget was too tight to pay for the construction of the new network of local streets to serve those houses.

To break the impasse, the real-estate company offered to build the new streets itself and apportion the cost among the purchase prices of the houses it hoped to sell. After five lengthy meetings of the town council, this arrangement seemed on the verge of approval.

But then came the question of maintaining the new streets. The town wasn't sure whether it could fit this extra cost into its tight budget. So the real-estate company offered to take on this responsibility. It would maintain the streets as well as build them and charge each homeowner a monthly service fee to cover the cost. In return, the town would levy lower real-estate taxes on the houses.

After several months, the proposal's financial advantages became so compelling that the town council voted five to three to pass it. The result is that Tom gets his driveway plowed free of snow each time his street is plowed, even though his total "home ownership costs" are about the same as mine.

It turns out that private ownership and operation of roads and other transportation facilities was common practice in the United States during much of the nineteenth century. What finally tilted the balance the other way for roads was the passage of the Federal Aid Road Act in 1916. That set the pattern for government—and government alone—to build, own, operate and maintain the rapidly growing networks of roads that spread across the nation during the next sixty years.

WHO PAYS AND HOW?

There has never been a shortage of ideas about how to pay for the nation's roads. At one end of the spectrum are people who think we should pay for them the way we pay for movie seats. Charge motorists an "admission fee" each time they use a road, they insist. This usually takes the form of a toll collected at some sort of barrier where drivers must stop and pay before proceeding. But today's technology can eliminate the stop and pay nuisance. It can even make it a simple matter to charge higher tolls at times when travel demand is high and lower tolls when demand is low, just as we do for movie tickets.

Many American bridges and tunnels charge tolls. A number of state turnpikes built before 1955 have used tolls to pay off their construction bonds and cover operating costs. The concept has an elegant simplicity that can be very appealing. The amount a driver pays in the course of a year depends on how often he uses the road. He makes his own decisions about whether the benefits of using the road for any particular trip are worth the admission fee.

At the other end of the spectrum are people who argue that the benefits provided by roads aren't confined only to drivers. By making it possible for people to travel (to work, to buy things, to deliver goods and services, to call on customers), roads provide benefits that flow through society in many complex ways. The ultimate measure of these benefits is the higher level of economic activity generated by so many people going so many places to do so many things. Therefore, the logic runs, society as a whole should pay for roads out of the general tax revenues collected by various levels of government.

Between these extremes lies a mixed bag of payment mechanisms. Most common are the various sales taxes on gasoline, tires and other things drivers have to buy. The more a person drives in the course of a year, the more of these things he buys and therefore the more he pays in sales taxes—which means each driver pays for roads roughly in proportion to the number of miles he drives in a year. This is a kind of indirect toll mechanism for roads in general, although it provides no incentive for a driver to use any particular road for his trip or to make that trip at any particular time of day. This method works best when the taxes collected are used exclusively to operate and maintain roads—which isn't always the case.

Over the years, the government's imposition of these payment mechanisms has helped lead us to the unconscious belief that roads have to be a purely government responsibility, like police and fire protection. But there is no iron law in physics or economics that this must be so.

A toll road can be built and operated by a private company just as easily as by a government turnpike authority. In fact, many toll-collecting turnpike authorities are actually profit-seeking corporations that happen to be owned by their home states rather than by private investors. But the choice of where ownership should lie is really an arbitrary one. The state of California realized this when it franchised a private company to build and operate a new toll road in the median strip of an overcrowded Southern California freeway.

In the same vein, the general tax revenues a government spends on roads can just as easily fund contract payments to private companies that provide various kinds of road services. Ditto the sales-tax revenues collected from drivers. A number of state and local transportation departments have already begun doing this.

In other words, it turns out that we have a surprising amount of leeway to decide whether roads should be operated by the public or the private sector. This means we can make decisions based on our judgment about who can deliver the best services for the least cost. We can even make these determinations on a road-by-road basis, just as my town did when it approved that plan to let the private real-estate company build and operate the local streets in Tom's neighborhood.

What influences our judgment in making these decisions? Sometimes it can be our personal experiences. If we live in a low-income neighborhood where we depend heavily on government services and where local retail stores seem to charge us ever-higher prices for poor-quality goods, we may feel that government is the "more reliable" service provider. If we live in a high-income neighborhood with its own private security service and send our children to private schools, we may be more comfortable buying whatever we need from private companies.

Most of us, however, are neither very poor nor very rich. If we bother to think at all about the question of who should provide which services (as I do when I commute on that decrepit highway), it's usually in terms of who seems able to do the best job. Academics like to steer this discussion into the realm of philosophy by talking about such concepts as "natural monopolies," "economic externalities," and "public versus the private goods." All of which ignore what may be a more pragmatic question: Has our growing mood of disenchantment with the apparent shortcomings of government increased our willingness to rely more on private companies to provide the public services we need?

If it has, we ought to ask ourselves how profit-seeking, tax-paying private companies may be able to deliver better services at less cost than not-for-profit, tax-free public agencies. Privatization advocates cite a number of reasons. The most important are probably better management, greater operating efficiency, absence of conflicting goals, and easier access to capital.

Better Management

An important key to efficient service delivery is smart management. This requires service managers who are bright, well trained, and strongly motivated. The simplest way to attract and hold on to such managers is to pay them better than what the competition does.

This is where public agencies are at a disadvantage. The salaries of elected officials tend to impose an artificial ceiling on how much public-agency managers can be paid, which is typically lower than what their counterparts in the private sector make. While this ignores the realities of the marketplace, elected officials are reluctant to raise these ceilings by advocating higher salaries for themselves, because they believe this "looks bad" to voters.

The people of my town and the rest of the American public seem to be entirely in sympathy with this salary constraint. They accept with unquestioning faith the strange idea that managers of public agencies should be paid considerably less than managers of private companies. The result is that the nation's best managers flock to higher-paying jobs in private companies, since public agencies are unable to compete for them in the marketplace.

But could we expect our favorite professional football team to make it to the Super Bowl if it insisted on paying its coaching staff less than what the competition did? The same principle holds true for the "coaching staffs" responsible for public services. Few professional managers are independently wealthy. Their choices about where to work have to reflect the financial realities of supporting their families and providing for their children's education. Inevitably, they have to work where they can earn the highest salaries. For the best managers, this is rarely in the public sector.

Given our peculiarly American attitude toward government and civil servants, there seems little likelihood that this salary constraint will ever disappear. So the only practical way to assure consistently high quality management of public services may be to turn at least some of them over to private companies who are willing and able to pay top dollar for ace managers.

Greater Operating Efficiency

Efficiency means getting more bang for the buck. Part of this has to do with good management. But sharp managers have to be free to exploit every opportunity for achieving greater efficiency, which the constraints on public agencies often prevent.

For example, public agencies must lean over backward to avoid the appearance of fraud or favoritism in their purchases of supplies and services. This can lead to elaborate bidding procedures for purchase contracts, lengthy audits before bills are paid, and endless amounts of time-consuming paperwork. Public-agency managers rarely have the option of simply picking up the phone and choosing a supplier based on timely delivery, quality products, and attractive prices. They're usually restricted to a small group of suppliers who have mastered the intricacies of government contracts, are willing to dot every "i" and cross every "t" (often several times), and wait months to be paid. All of which means higher costs that get built into the price of every purchase contract.

Avoiding "waste, fraud and abuse" is the standard litany that guides public-agency purchases. No recognition is given to the pragmatic trade-offs that must often be made between these three demons. The standard assumption seems to be that the public would rather waste countless extra dollars to avoid any possibility of losing a single dollar to fraud or abuse. And no public-agency manager can be expected to lay his career on the line to exploit more-efficient trade-offs.

The bottom-line focus of most private companies, however, forces them to live in the real world, where practical results are what count, not elaborate procedures. That's why managers working in such a culture can exploit the benefits of just-in-time inventory management, economies of scale in purchase orders, and meaningful measures of supplier performance to make their operations more efficient.

Great efficiency in service delivery also means being able to aggressively exploit the potential of new technology to streamline production and reduce costs. But there's always some risk in being among the first to embrace new technology. Managers of private companies are paid to accept and manage risk, since that's what business is all about, while public-sector managers are paid to avoid risk. This is why public agencies are usually the last to adopt modern technology. They prefer not to "risk the public's money" until a better technology has been so totally proven in practice and completely debugged that it's often obsolete.

Absence of Conflicting Goals

In theory, a public agency should have only one goal—to deliver effective services with maximum efficiency. In practice, however, this can often be complicated by other goals, such as providing various social benefits that can take the form of more entry-level jobs for young people, higher community standards for blue-collar wages and benefits, improved opportunities for women and minorities, and so forth. But when these goals conflict with an agency's service-delivery goal, they can end up causing both public services and social benefits to cost more than they should.

These conflicts are easier to avoid when a private company provides public services under a government contract. In such cases, any ancillary social-welfare goals would have to be explicitly stated in the contract, and this tends to discourage attempts to make them part of the service-delivery package.

But what about the potential for conflict between effective service delivery and a private firm's "need to make a profit"? Of course, management guru Peter Drucker has always argued that "creating customers" should be the main goal for any private firm, with profit simply being one of the costs it must cover (just like paying employees and suppliers). This sounds like an ideal approach for managing public services. Unfortunately, there's little evidence that it's being widely practiced in the public sector.

As it turns out, the marketplace can often provide a practical solution to these potentials for conflicting goals. This can happen when public agencies and private firms are allowed to compete on price for contracts to deliver public services.

In a surprising number of cases, such open competition has enabled public agencies to become as cost-efficient as the best private firms. In other cases, private firms have found it in their best interests to shave their profit targets somewhat to win service contracts. Either result simply means that our willingness to let the marketplace help us choose a service provider has created the kind of environment that leads to better public services at less cost. Which is, after all, what we're supposedly after.

Easier Access to Low-cost Capital

Most public services require investments in capital facilities. Roads are an obvious example. First we have to build them, which takes capital dollars. After a certain number of years, they wear out and we have to rebuild them. This also takes capital dollars. Along the way, the travel demands on them may become so great that we have to expand them, which requires still more capital dollars. From one perspective (admittedly a rather narrow one), the whole issue of public versus private operation can be seen as a question of who is best able to raise the necessary capital at the least cost.

State and local governments are subject to all kinds of legal restrictions on the amount of debt they can issue and for what purpose. In some cases, debt issues must be approved in advance by voters, and such referendums can often be difficult to pass. In other cases, elaborate formulas may be built into state constitutions that link the maximum amount of debt a county or city may have outstanding to some arbitrary fiscal measure—such as the value of its taxable real estate. The end result can limit the ability of state and local governments to raise enough capital to build the roads and other public facilities that today's economy requires.

Some states have tried to get around these debt-issuing restrictions by backdoor means. For example, a state turnpike authority may be able to issue whatever amount of debt that its toll revenues can support. So a toll increase can open the door for additional borrowing. An amendment to its charter—or simply a less rigid interpretation of this charter —can let the authority use these new funds to build or expand toll-free state roads that it claims (with a perfectly straight face) will feed additional traffic to its toll road. Even more ingenious schemes are possible. But their overall effect tends to be marginal.

Over the years, the federal government has attempted to fill the capital gap through its various transportation assistance programs. Current law allows the federal government to fund up to 80 percent of the cost of an "eligible transportation project." But this isn't nearly as open-ended as it sounds. The annual federal appropriation process, coupled with Washington's new balanced-budget fetish, means that there's never enough federal aid to go around. And there's no reason to expect this to get any better.

Traditional finance theory argues that all debt should be self-financing. That is, the capital facilities it creates should generate a new income stream to cover interest and principal payments by a safe margin. But the flip side of this market-oriented restriction means that it's fiscally prudent to issue increasing amounts of debt to exploit whatever new economic opportunities are able to satisfy the self-financing mandate.

The world is awash with capital-seeking investment opportunities. Large corporations, unencumbered by the arbitrary fund-raising restrictions that hamstring government, have access to this capital so long as they can demonstrate their ability to put it to work in ways that generate enough new revenue. In other words, the future is what counts, not the past or the present.

Many corporations have relied on this open access to capital to finance the development of new products that they expect to sell at a profit. What if some of those products could be the roads and other public goods traditionally regarded as the exclusive preserve of government?

Could this help bring an end to our artificially created inability to afford the infrastructure on which our future prosperity depends?

That's how Tom's neighborhood got the local streets that our town couldn't afford to build and maintain—streets without which none of those new houses could have been built in the first place.

Access to capital is only part of the story, however. The rest of it concerns the cost of capital, which we want to keep as low as possible. A state or municipal government can raise capital funds for its public agencies by issuing bends whose interest payments are not subject to income taxes. All else being equal, this means that these tax-free bonds enjoy lower interest rates than the taxable bonds issued by private corporations.

On the face of it, this would seem to give public agencies a capital cost edge. But in many cases, the reality may turn out to be less simple and more interesting. It happens that the federal government provides two attractive subsidies to private companies that are not available to public agencies. Under the right circumstances, this can have a significant impact in reducing a private company's true cost of capital.

One of these subsidies is the full deductibility of interest payments in determining the company's taxable income. For many profitable companies, this can result in out-of-pocket interest costs that are competitive with tax-free debt. Since public agencies have no income-tax liabilities, their nominal interest costs are identical to their out-of-pocket costs. And out-of-pocket costs are what count.

The second federal subsidy involves how private companies are allowed to deduct "depreciation"—which is the prorated cost of a capital facility over the years of its useful life. Even though each year's theoretical depreciation cost doesn't actually mean that the company had to lay out any cash, the IRS lets the company pretend it did when it computes its taxable income.

In addition, the IRS lets the company use a "tax depreciation" schedule for certain capital facilities that is more generous than normal "accounting depreciation" by allowing larger deductions in the early years of the facility's life. Since all deductions for depreciation mean lower taxable profits, the company's actual profit ("return on invested capital") turns out to be considerably higher than the theoretical profit it must report to the IRS.

In the words of the accounting profession, these tax deductions for interest and depreciation flow directly to a firm's bottom line. And their practical effect can sometimes reduce the company's cost of capital to levels that public agencies can't match. In certain cases, large corporations with solid credit status can obtain capital funds for as little as their same cost to a public agency.

CAN IT REALLY WORK?

Can private operation really work on a large scale? Not unless we change our thinking about that Chinese Wall we've erected between the public and private sectors.

In the largest sense, we must learn to see government and private companies as natural partners—not competitors—in providing the wide range of public services our society needs if it is to prosper in the future. Government has the obligation (and the power) to set service-delivery standards that focus on the welfare of society as a whole rather than the narrowly defined economic welfare of any particular company. The private sector has the ability to attract gifted managers who can meet these standards efficiently—and to attract the capital funds needed for the modern infrastructure that is essential for effective service delivery.

Once we accept the implications of this natural partnership, we can move on to the admittedly tedious process of dismantling the numerous legal and administrative barriers that dot the public service landscape. The results may surprise us, probably in some very pleasant ways.

Private operation of the streets in Tom's neighborhood has worked so well that no one even talks about it anymore. Public operation of that highway on which I must commute every day doesn't seem to work for anyone. Except, perhaps, for local newspapers in search of convenient stories about "government's increasing dysfunction."

Is there a lesson here?

Emerging Opportunities for Public–Private Partnerships in Highway Development

Privatization is often viewed as an arrangement under which the private sector becomes involved in the financing, design, construction, ownership or operation of public facilities or services. Another term conveying the same meaning is public–private partnerships. The underlying concept is that both the public and private sectors can find mutual interest in cooperating to provide services and facilities.

Primary Goal: Cost Savings

If the public sector had to list ten reasons for the renewed interest in public–private partnerships, the first nine would relate to cost savings. Indeed, most of the 1,086 cities and counties that responded to a 1987 survey cited cost savings as the leading advantage of privatization. A survey conducted by the Council of State Governments in 1993 reported the same result, with cost savings cited as the top reported reason for privatization.

Such arrangements are more common than most people realize. Governments, for example, have always used private firms to prepare engineering and architectural designs for public buildings. Essential public services such as electricity, gas and telephone communication have traditionally been provided by private firms that function as regulated monopolies.

Adapted from Joseph M. Giglio, "Emerging Opportunities for Public–Private Partnerships in Highway Development," in Restructuring State and Local Services: Ideas, Proposals, and Experiments, *ed. Arnold H. Raphaelson, 45–63 (Westport, CT: Praeger Publishers, 1998).*

What is new is the concept of having private firms operate public libraries, prisons, sanitation services, toll roads or other functions normally associated with public agencies. Contemporary privatization represents a collaborative effect between the public and private sectors, sharing risks, rewards and responsibilities.

After more than a decade of rhetoric, interest in privatization opportunities is increasing, partially driven by the fiscal condition of government and a demand for services exceeding available resources. The assumption behind this increased interest is that private firms can often deliver these services for less cost. Although statistical evidence is sketchy, several studies indicate that private-sector involvement can result in substantial cost reduction. For example, in developed countries, transferring bus services from public to private operation has resulted in savings of up to 20 percent.

FACTORS THAT RESULT IN COST SAVING

How is this possible, since a private firm must pay taxes and make a profit—two burdens that public agencies do not have? The usual answer to this question rests on the expectation of something called private-sector efficiency. This vagueness does privatization a disservice and can often impose unnecessary barriers to its implementation.

There are at least four factors that can work in a private firm's favor:

1. ***Higher private sector salary and incentives.*** A public agency's civil service salary structure and inability to offer incentives such as stock options may make it impossible to attract a sufficient number of talented managers. Effective management is a major factor in being able to deliver quality service at low cost. Poor management leads to waste and inefficiency, which degrades service quality and boosts costs.

2. ***Faster procurement process.*** A public agency's procedure for purchasing materials may be constrained by a variety of regulations that were originally implemented to prevent fraud. But the practical effect can complicate the agency's inventory management. Purchases can take an inordinately long time to complete, increasing the danger of running out of supplies. To guard against such outages, the agency may overbuy, tying up its working capital and foregoing any possibility of exploiting the cost-saving benefits of just-

in-time inventory management. This can lead suppliers to charge higher prices to compensate for payment delays. Since private firms are not subject to these regulations, their purchasing and payment procedures can be simpler, faster and less costly.

3. ***Economies of scale.*** By providing the same services to a number of different cities, a private contractor can develop both market clout and specialized expertise in a particular area. This can be particularly critical for high-technology services. For example, if a city wanted to purchase advanced data-analysis equipment for its own needs, it would be charged a higher rate than a contractor who purchased the same equipment in bulk.
4. ***Less restrictive work rules.*** A public agency may be saddled with labor work rules that prevent the implementation of new and more efficient procedures. A private firm that does not have a long contractual history with the labor unions involved may be in a better position to negotiate work rules that enhance efficiency.
5. ***Availability of tax deductions.*** Finally, there are two kinds of automatic federal subsidies available to private firms that are not available to public agencies: tax deductions for accelerated depreciation on capital equipment and for interest payments on borrowed funds. Of course, federal taxes do not apply to public agencies. These tax benefits are not usually considered federal subsidies, but that is their practical effect. In some cases, they may enable a private firm to have a lower after-tax cost of capital than a public agency. Under certain circumstances the combined effect of these factors can enable a firm to deliver better services at a lower cost than is possible for a public agency.

Types of Public–Private Partnerships

Given the broad definition of privatization, the public sector can select from a variety of structures in collaborating with the private sector, like government contracts. There are no simple models, but rather a set of techniques that allocate responsibilities of ownership, risk, operation and oversight among public and private parties.

Asset Sale

One of the privatization techniques involves sales of assets. In one version, the government sells the asset to a private buyer, who then operates the asset for a public purpose. The sale by the federal government of Conrail in 1987 is an example of the sale of an enterprise as a functioning unit through a public stock offering.

The area of asset sales has been a major source of misrepresentation about the nature of privatization. Consider the case of a city government facing a budget deficit that is against state law. City officials are usually reluctant to raise property taxes or cut services; municipal elections are often less than a year away. Then a local real-estate developer may offer a solution. The private firm will buy the city's antiquated public library building for a flat sum that is large enough to balance the budget. The private developer will finance this purchase through a loan from a local bank. Then the private firm will lease the library building back to the city for annual rental payments equal to the debt service on the bank loan.

What has been accomplished? The city has balanced its budget for the upcoming year. The developer has helped officials out of difficult situation, which may not be forgotten the next time the private developer seeks a zoning change. The bank has a new loan, whose safety is secured by the city's taxing power. Taxes are not raised, nor are services cut.

But the rundown library building is still a maintenance burden for the city's government, which must also continue to operate library services. And taxpayers have been saddled with a new series of annual costs to meet the rental payments, which does nothing to improve services. Politicians are sometimes quick to embrace simple answers to complex issues that make them look good and feel good, but in fact, make a bad situation worse. Privatization should not be a financial "fig leaf" to buy politicians a few years to figure out what to do next and then hand the problem off to their successors.

Contracting Services

The contracting out of services is another technique. The public sector contracts with a private firm to provide a specific service, such as garbage collections, instead of producing the service itself. Ownership still resides with the public sector. Under this technique, the public officials set the policy goals, and the private firm implements them. The goal of contracting for services is to achieve efficient service delivered for the price government is willing to pay. This technique is widespread and is growing at the state and local levels of government.

A variation of contracting out is franchising, under which governments award an exclusive right to deliver a public service or good to a private firm. Among the services commonly franchised are water, electric, gas, telephone and cable television.

Construction and Operation Arrangements

In another privatization technique, a private firm or consortium may build or acquire a facility, such as a toll road, and then own and operate it to serve the general public. There are variations of this public–private structure based on the sharing of the responsibilities, risks and rewards of each party.

The private firm or consortium may build and operate the asset under the so-called Build–Transfer–Operate (BTO) model for a limited time period, or transfer it to the public sponsor immediately after construction. This model may be used for the construction of new toll-ways with the financing, construction and operating risks born by the private sector.

Alternately, a private firm or consortium may receive a concession to build, finance, own and operate an asset for an extended time period, after which the asset is transferred to the public sponsor, the so-called Build–Operate–Transfer (BOT) model. The government may get out of the project entirely and relinquish the operation of a service to a private firm, relying on its regulatory powers to oversee operations. There are variations on these models, reflecting different collaborative approaches. As government finds itself faced with more demand for service and without the financial resources to provide the assets, the private sector can become a source of much-needed capital.

Evaluating Public–Private Partnerships

Before moving on to considering privatization opportunities in highway development, it would be useful to evaluate public–privatization arrangements from the public sector's perspective. What is needed is a litmus test to determine whether a particular privatization arrangement is legitimate and beneficial. It should include the following criteria:

1. ***The transaction must be legal.*** This may be obvious, but many state and local governments have laws that effectively preclude public–private partnerships. The sponsor of a transaction must be sure of the underlying legislation; in many instances, new legislation is required.
2. ***The transaction must be smart, from the government's perspective.*** It must have a valid—and hopefully quantifiable—business purpose. State laws may allow the proceeds of asset sales to be recognized as current revenue for budgeting purposes. But their use to fund deficits may be little more than a budgetary trick. To be a legitimate transaction, the privatization must provide a long-term benefit for the government. In many cases, that benefit will be the reduced cost of providing a service to the public.
3. ***The cost implications of the transaction must be understood in advance.*** This is especially critical when cost savings or even revenue improvements are not the major benefit from a transaction.
4. ***The transaction contract must spell out key provisions and performance standards.*** The formal contract is required to ensure that the private partner maintains the minimum level of service required by the public agency.
5. ***The government must have mechanisms in place to monitor the performance of the private firm.*** The government cannot abdicate its responsibility to provide a service that it has a legal obligation to provide.
6. ***The transaction should be subject to competition.*** This competition can either be among private-sector firms or between the public and the private sectors. At times, even the potential for public–private partnerships has resulted in competition and increased productivity in the public sector workforce, obviating the need for the privatization transaction.
7. ***Displacement or layoffs of public employees should be minimized as much as possible.*** Privatization is a tricky term. Unions may perceive it as a code word used by elected officials not just to invoke a collaborative effort to promote good government but—in the same way crosses are whipped out to scare off vampires—to threaten their existence.

Early and active union involvement in such arrangements is more helpful than harmful. Indeed, despite union fears, widespread layoffs are not necessarily a consequence of privatization. A 1989 study by the National Commission on Employment Policy, a research arm of the U.S. Labor Department, reviewed thirty-four city and county privatizations and concluded that only 7 percent of affected workers were laid off.

8. ***Governments must be prepared for the public relations difficulties often associated with public–private partnerships.*** Any government wishing to privatize public services faces the virtual inevitability of periodic and publicized scandals. Stories about public scandals attract audiences. Consequently, editors and producers urge their reporters to ferret out real or apparent incidents of improper behavior in dealings between the public sector and private firms. Many such incidents may involve no violations of the law. This scandal potential is inherent in all public–private partnership arrangements, and may be the most challenging barrier to its widespread application.

Public–private partnerships are an important public-policy technique that enables governments to deal with the current cost-revenue-service squeeze. We must hold our policy makers to a set of criteria and guidelines to ensure that they do more than balance annual budgets; rather, that they also consider the impact of their transactions on the taxpayers of today and tomorrow.

Demand for Public–Private Partnerships in Transportation

Historically, government agencies have made extensive use of the private sector for construction, maintenance, design and engineering services. A 1992 survey found that 85 percent of state governments had contracted out at least some functions related to highways.

Before recent federal legislation and state initiatives, however, public–private arrangements that involved private-sector ownership or operation of highway facilities have been rare, compared to other infrastructure sectors. For example, private ownership and operation of drinking-water facilities is commonplace, yet private ownership and operation of toll-road facilities is uncommon.

But at all levels of government in the United States, the demand for public funds for the construction and improvement of transportation infrastructure is greater than the available funds. The federal government has been struggling to bring the federal deficit under control while pursuing healthcare reform and other policy goals. State and local governments confront similar budgetary pressures. The nation has been suffering from a capital under-investment in infrastructure systems for more than two decades.

Decreased Funding For Infrastructure

For the past twenty years, the country has steadily devoted less of our resources to infrastructure. Only 2.6 percent of the gross national product in 1990 was accounted for by infrastructure, compared to 3.7 percent in the early 1960s. When these numbers are adjusted for depreciation, we have been investing less than 0.5 percent of our gross national product for the last fifteen years. Of the developed world, only Great Britain comes close to this dismal record. Japan invests 4 to 6 percent of its gross national product in public projects. For some reason, we virtually stopped net capital investment in the public sector in the late 1960s and 1970s. Even with more than a trillion dollars of net value in public works, how long can we live off the past?

In some areas we have actually begun to disinvest. The net asset value of our roads, bridges and streets is lower than it was ten years ago. But work trips are not down. Ton-miles of cargo hauled are not down. Daily aggravation is not down. The decline in investment has been so dramatic that a visitor from outer space would have to conclude that the people of the United States have been blessed with a dramatic new technology about to render the old modes of transportation obsolete. After all, why would an educated, powerful country stop investing in its future?

Decline in Productivity

Dr. David Aschauer, formerly of the Federal Reserve Bank of Chicago, has developed statistical models that show a direct link between the decline in public investment and the decline in total productivity. Indeed, he has shown that more than half of the country's drop in economic productivity can be traced to the drop in public-works investment. His studies show that the most significant single factor in the recent decline in U.S. productivity has been the drop in public-works investment. Still further, his research demonstrates that there is a robust, positive linkage between public capital—particularly infrastructure capital—and private-sector productivity. He also found a long-run positive relationship between public, nonmilitary investment and business investment.[1] Other scholars have expanded upon and largely confirmed his research.

The decline in public investment that began in the last half of the 1960s is a significant cause of the inferior economic performance of the ensuing two decades. Over this time, the United States experienced a lower growth in productivity in the private business sector (1.4% annually rather than the 2.8% of the 1950s and 1960s), a diminished rate of return on non-financial capital (7.9 percent vs. 10.7 percent) and a drop in the growth rate for the private stock of plant and machinery (3.1 percent annual growth vs. 3.8 percent). The National Council on Public Works Improvement rated the nation's infrastructure a "C." More specifically, highways received a grade of "C+." The Federal Highway Administration's biannual report to Congress, *Status and Conditions of the Nation's Highways*, documents the physical problems faced by our roads and bridges.[2]

The U.S. transportation network is the largest, most complex in the world. An efficient transportation system is essential for economic growth. Almost one of every five dollars spent on goods and services in the U.S. economy is spent on transportation products and services. Our transportation infrastructure is not only deteriorating from aging facilities, but capital under-investment in facilities is also overburdening the current transportation system. Recent reports indicate that between 1998 and 2002, highway expenses to maintain current conditions and capacity will reach $263.7 billion, with an additional $93.8 billion required for improving current conditions. The total of $357.5 billion in highway needs will exceed current expenditures by more than $42 billion.

In recognition of the importance of transportation infrastructure for economic prosperity, many public policy makers have concluded that highways are not the sole responsibility of the government. As a result, government transportation agencies are turning to the private sector to expand the pool of available capital funds for investment in transportation infrastructure. While private-sector involvement has been growing in all areas of infrastructure, surface transportation may be a particularly fertile area to attract private-sector capital to bridge the funding shortfalls resulting from the growth in transportation needs and the decline in public funding sources. Highway projects supported by tolls are particularly attractive investment opportunities because the identifiable, separate revenue stream supports a financial return to the private sector. The linkage between the use and cost for the system promotes rational consumer and investor behavior.

Evolution of Highway Financing

Recent federal transportation legislation was designed to provide a strong federal partnership in the development of the nation's transportation infrastructure. While the legislation has provided states with more flexibility than funds, a number of provisions allow states to attract private capital as an antidote to the federal budget constraints. It also has allowed them to more effectively leverage federal dollars with both public and private-sector resources in order to expand the total available for transportation investment. One key feature of this new legislation is the new federal policy of allowing state and local governments to utilize federal funds on projects where tolls can be charged—a major shift in the historic relationship between federal and state support. Using toll revenues to match and leverage federal aid provides states with the ability to expand the financial resources available for transportation infrastructure projects in a manner that was not possible before.

Although a federal role in highway construction dates back to the early days of the republic, the modern highway program has its beginnings in the Federal Aid Road Act of 1916. This act established many of the basic provisions of federal highway policy that are still in effect. The most important of these provisions is the federal–state partnership whereby states retain ownership of the roads and the responsibility for their construction and maintenance, while the federal government provides financial aid to the states for construction in the form of federal-aid highway funds. It is important to note that federal-aid highway funds are not termed "grants" because the funds are regarded as reimbursement to the states from the gas taxes charged to state residents. These funds are apportioned according to formulas based on such factors as area, population and road mileage. This historic division of effort—federal support of construction, and state and local responsibility for care and maintenance—was in place until the late 1960s, when the deteriorating condition of existing roads led to gradual change in federal regulations to permit the use of federal funds for major repairs.

Beginning with the Federal Aid Road Act of 1916 and the Federal Highway Act of 1921, federal policy discouraged toll financing. The acts specified that roads and bridges constructed with federal funds were to be free of tolls. Congress, however, granted a number of exceptions to the no-toll policy over time. These exceptions permitted the use of federal money to construct toll bridges and tunnels and their approaches on the federal highway network. Federal policy required that the revenue generated by toll collection be used to pay solely for the construction, maintenance and operation of the facility. This policy was intended to prevent cross-subsidies—taking toll revenues, for example, from one facility and converting them to another facility. Also, federal policy required that tolls be removed from federal aid facilities as soon as the original construction bonds were retired. Congress did, however, grant several extensions to deadlines for removal of tolls on federal aid facilities.

The demand for more and better roads grew rapidly following World War II. To help alleviate the pent-up demand, state and local governments began placing greater emphasis on toll facilities, and serious consideration was given to a national system of limited-access toll highways. This interest in toll roads, however, was greatly reduced with passage of the Federal Aid Highway Act of 1956. This Act provided 90-percent federal financing for more than 41,000 miles of interstate and defense highways, an increase in federal funds for roads on the existing federal aid system, and the creation of the Highway Trust Fund to provide a continuous source of funding. These features made reliance on federal resources for highway construction increasingly attractive to the states. Subsequent increases in motor fuel taxes and vehicle taxes, enacted in 1959 and 1982, continued to make reliance on federal funding attractive.

Congress also provided for some toll roads built prior to 1956 to be included in the interstate system, as long as the toll facilities on these portions were removed at the earliest practical opportunity. In addition to provisions for certain existing toll facilities, Congress would create exceptions on a case-by-case basis. Many of these exceptions required that all federal funds be repaid, but the payment terms were extremely generous because there was no interest charge on the repayment program. In addition, these funds were credited back to the state's unused balance of federal-aid funds. It was apparent that this policy created strong economic incentives for states to retain toll facilities. Many more such exemptions were granted under the auspices of the Surface Transportation Act of 1978, which allowed federal revenue to be spent on resurfacing, restoration, rehabilitation and reconstruction—"4R" funds—on the toll segments of the interstate system. This act represented a departure from the early federal policy of strict opposition to toll roads.

The federal government continued to move away from a strict policy of opposition to toll roads with the passage of the Surface Transportation and Uniform Relocation Act of 1987. This act provided for federal participation in seven pilot toll projects. Under this legislation, the federal government provided a 35-percent subsidy of construction or reconstruction costs of the seven projects in an effort to more effectively finance these needed facilities.

In sum, by 1987, the federal policy opposing toll roads had been gradually eroded, and toll bridges and tunnels had been constructed with federal revenue. Consequently, roads connected to toll facilities were constructed with federal revenue. In 1978 toll facilities became eligible for federal 4R revenue, and in 1987 Congress passed a five-year highway program that provided federal revenue for the construction or reconstruction of the seven pilot toll roads. Thus, for the first time since federal aid to highways began in 1916, states were allowed to put tolls on existing and new federally funded bridges, tunnels and roads other than interstates. Although to date all but two of the toll roads in the United States have been publicly owned (the Dulles Greenway in Virginia and SR-91 in California), the increased flexibility in using tolls on federal-aid projects widens many potential opportunities for public–private partnerships in financing, operating and owning toll roads.

RECENT DEVELOPMENTS IN FUNDING PUBLIC–PRIVATE PARTNERSHIPS FOR HIGHWAYS

In 1991, the federal government passed the comprehensive Intermodal Surface Transportation Efficiency Act (ISTEA), designed to provide a strong federal partnership in the development of the nation's transportation infrastructure. The centerpiece of the legislation was an expanded level of flexibility in designing projects, allocating resources and developing funding mechanisms. Part of this flexibility stems from an increasing ability to enter into partnerships with the private sector. State governments were put in a unique position to play a leading role in using this flexibility to dramatically expand the funds that are available for transportation infrastructure investment. ISTEA provided an opportunity to reshape the surface transportation agenda, and it signified the end of an era that began in 1956 when the Interstate Highway Program was originally authorized.

ISTEA was the first in a series of developments that have increased the ability of states to finance transportation investment in new ways. In a 1994 Executive Order, "Principles for Federal Infrastructure Investments," the Clinton administration directed federal agencies to explore new ways of encouraging private participation in infrastructure investment. In April 1994, in response to the innovative financing opportunities available under ISTEA and to this Executive Order, the Federal Highway Administration (FHWA) used its authority under current law to launch a test and evaluation program (TE-045) that asked states to identify and propose new financial strategies to leverage federal dollars to increase investment in transportation infrastructure.

Under this project, states have proposed regulatory and policy changes to facilitate innovative project financing. To date, more than sixty projects from over thirty states have been accepted for a total construction value of over $5 billion. States submitted five categories of proposals to FHWA under TE-045:

1. Innovative management of federal funds
2. Bonds and other forms of debt financing
3. Expanded matching opportunities
4. ISTEA's Section 1012 loans
5. Innovative income generation

Using the experience gained from the TE-045 test and evaluation projects, the National Highway System Act of 1995 codified several innovative finance concepts. They include permitting states to use federal-aid funding to pay bond principal, interest, issuance costs and insurance on ISTEA-eligible projects; allowing toll roads to be funded with an 80-percent federal share; removing limitations on advance construction; establishing a State Infrastructure Bank pilot program; expanding Section 1012 of ISTEA to allow loans to non-tolled projects with dedicated revenue streams; permitting below-market interest rates; allowing private-sector donations of right-of-way and in-kind services to count against the local share; and expanding the eligibility of toll roads for up to 80-percent federal-aid funding.

Finally, the reauthorization of ISTEA expected in 1997 will likely lead to further developments that enhance state ability to enter into public–private arrangements for highway development. Here again, the object of these statutory and policy changes is to enable states to attract private capital as an antidote to federal budget constraints, as well as to more effectively leverage federal dollars with public and private-sector resources to expand the total amount of funds available for transportation-infrastructure investment.

The Changing Highway Financing Landscape

These legislative and policy developments are consistent with the changing demands of the nation's transportation system. Economic changes, including growth in the global economy and shifts in corporate structure and ways of managing operations, have created a new set of demands on transportation infrastructure. Societal and demographic shifts, such as increases in dual-wage-earner families, an aging population and a shift in employment and residences to suburban and exurban locations have altered the use of our highway landscape. An increasing awareness of environmental problems and a link between environment and transportation have complicated the traditional transportation focus on mobility as a primary objective. Also, new technology, including those technologies that fall under the Intelligent Transportation System (ITS) rubric, have created many new capital financing opportunities for addressing transportation needs.

Moreover, the legal and regulatory framework for transportation decision-making has changed markedly in the last several years. ISTEA and the Clean Air Act Amendments of 1990 have combined to create a new framework for transportation planning and finance. Together, these two new laws are reshaping the operating context for state departments of transportation, metropolitan planning organizations and others, both by increasing financial flexibility and by altering the decision-making process and the criteria applied to transportation investment decisions.

Presumably, state transportation departments and metropolitan planning organizations will work together to make all the resource-allocation decisions of consequence. Yet integrating transportation planning and financing will be accomplished against great odds. Even in the best of times, efforts to tie planning and financing together tend to totter. The two groups need each other but have different agendas.

These new demands and changing legal and regulatory movements have led to a search for new innovative solutions to transportation financing issues. Rather than simply building more highways, transportation professionals are seeking more creative ways to address multi-dimensional transportation problems such as convenience, safety and pollution. Increasingly, they are beginning to consider new technologies, including ITS. ISTEA's liberal definition of flexibility means that federal funds are available for investment in high-technology electronics; for pavement, bridge painting and deck patching; for reconstruction, planning and research development; as well as for the start-up costs of traffic management and control. At the same time, as government budgets get tighter, transportation agencies are looking for new ways to finance important infrastructure projects. Many are considering State Infrastructure Banks as an alternative capital source to meet transportation demands in a fiscally constrained environment.

Increased Flexibility in Funding Categories

A key feature of ISTEA and other policy developments is the increased flexibility afforded state and local governments in using federal funds for transportation investment. In addition to creating a highly flexible new program (the Surface Transportation Program) which allows highway funds to be spent on transit and other programs, ISTEA permits up to 50 percent (and in certain cases 100 percent) of national highway system funds and up to 40 percent of bridge funds to be used for transit programs. ISTEA also provides significant incentives to state governments to involve the private sector in the ownership of toll facilities. In addition, the National Highway System Act of 1995 increased the allowable percentage of federal funding for toll roads to 80 percent, the same as most other federal-aid projects.

Increased flexibility in funding categories enable public–private partnerships by facilitating the blending of public and private funds, and by encouraging state transportation departments (DOTs) to experiment with new programs and sources of revenue. The flexibility also allows arrangements that had not even been considered before. For example, Missouri recently received approval under TE-045 to use a pledge of its future highway appropriations as credit enhancement for a project sponsored by a consortium of private railroads. The project is a fly-over (an above-grade crossing) that would eliminate a four-way rail-highway crossing causing significant highway congestion in Kansas City. While the state was reluctant to use highway funds directly for this private project, the congestion mitigation benefits were substantial, and the state had a strong vested interest in ensuring that the private consortium was able to attract capital, even without an existing credit rating. The availability of the apportionment's pledge will enable this project to be fully financed with private funds, assuming that no draw-down of the credit enhancement occurs.

Innovative Management of Federal Funds

Under TE-045, the Innovative Finance Research Initiative, states also requested the following innovations:

1. ***Phased funding***—allowing a state to begin construction before obligating the full amount to a project.
2. ***Tapering***—allowing the federal share to vary over the life of a project, as long as it does not exceed the maximum 80 percent.
3. ***Advance construction***—allowing states to seek future reimbursement for self-financed projects.
4. ***Partial conversion of advance construction***—allowing states to seek reimbursement for projects over a period of years, instead of all at once.

Though many states will use them solely to better manage public-sector resources, these techniques also have the potential to substantially increase a state's ability to engage in public–private projects. For example, if a state were interested in building a toll road that would ultimately be publicly owned, it could enter into a BTO arrangement with a private partner and seek future federal-aid reimbursement by making use of advance construction. The transportation agency would still be able to use federal-aid funds to complete the toll road, but would enjoy the advantages of faster and less expensive private-sector procurement and construction.

Flexible Funds Matching

States can use both flexible matches and Section 1044 toll credits to maximize the use of all available resources to finance transportation investments.

Flexible Match. Under TE-045, states can include the value of in-kind contributions, publicly donated rights-of-way (ROW), local government capital, private funds or credit for previously constructed projects in the local share. This provides a strong incentive to states to attract private roadway investment and to foster more public–private partnerships of all kinds. These partnerships could range from a group of business owners seeking a new interchange and who donate funds for ROW purchase, to a telecommunications provider who installed ITS services in exchange for access to highway ROW. In particular, this matching provision provides strong incentive for these kinds of "shared resources" arrangement, in which the public sector provides access to public ROW in exchange for cash or in-kind compensation. For example, Missouri sought matching credit for the value of ITS technology installed along a highway ROW as part of a shared resources arrangement. The private partner was permitted to use the ROW; the state received the ITS technology as compensation, and was also permitted to count the value of the ITS technology against the required local share for the project.

Section 1044 Investment Credits. Under Section 1044 of ISTEA, states can receive investment credits for past toll-road investments that count toward a nonfederal match in future years. To be eligible for Section 1044 investment credits, states must demonstrate a steady commitment to transportation investments via the "maintenance of effort" provision, which requires that they maintain a level of spending equivalent to an average of historical spending. Many states have had difficulty complying with this provision, often due to insufficient investment in only one of several prior years. As a result of TE-045, FHWA has changed its guidance and is now allowing states to choose among several tests, including the use of prospective spending commitments to meet this standard.

The availability of Section 1044 credits could solve a significant impediment to Section 1012 loans (discussed below). To make these loans, states are required to come up with a 20-percent match to the federal funds. In many states, the political climate might not allow for the use of state funds as a loan to private parties. If a state had Section 1044 credits from previous toll-road investments, it could meet the matching requirements for Section 1012 loans without using state funds.

Leveraging Tools

Two kinds of leveraging tools that assist in public–private partnerships have been created or enhanced by recent federal legislation: Section 1012 loans and State Infrastructure Banks (SIBs).

Section 1012 Loans. Under Section 1012 of ISTEA, states can loan federal funds to revenue-generating facilities such as toll highways, tunnels and bridges. Loans can be made to public, quasi-public or private project sponsors, may be subordinate to senior debt, can have an extended repayment period, and can use short-term interest rates. Section 1012 increases a state's flexibility in meeting its financing needs as well as its ability to stretch federal funds. Under the TE-045 initiative, FHWA received and may consider proposals allowing states to do the following:

- Offer assistance other than direct loans, such as credit enhancement.
- Use loan repayments to capitalize a revolving fund.
- Develop alternative repayment sources beyond toll revenues (e.g., dedicated gas-tax revenues, sales taxes, etc.).
- Make loans to projects that use ITS technology.

Section 1012 of ISTEA allows states to loan federal-aid highway funds to all eligible Title 23 projects—most highway and transit investments—under public or private control. In particular, Section 1012 of ISTEA allows state transportation departments to loan federal funds to public or private entities constructing toll facilities. Rather than using federal money to fund a toll facility project directly (i.e., reimburse project costs), the state may take the money and loan it to a private partner. The federal shares are the same as for straight toll financing, up to 50 percent for highways and 80 percent for bridges and tunnels. As toll revenues begin to be generated, the toll operator repays the state principal and interest. The state can then re-loan the funds to another transportation project. This program increases the number of projects the state can construct over time. Specifically, the loan provision operates as follows:

1. The loan is repayable to the slate transportation department beginning not more than five years from the date the facilities are opened to traffic.
2. There is no requirement on the method of amortization.
3. The loan term may be up to thirty years.
4. The debt is subordinate to other project debt.
5. Repayments can be re-loaned for any Title 23 transportation purposes.

States can use Section 1012 loans to provide start-up financing for toll roads and other privately sponsored projects. Because the terms of the loan permit delay of repayment until five years after project completion, the private partner will have the time to generate sufficient toll revenues for repayment. In the event that traffic does not materialize as rapidly as projected, leading to a default, the Section 1012 loan would be subordinate to other forms of debt. Under this scenario, a state might choose to permit the toll road to continue to operate even after default, in the hope that sufficient traffic will eventually be generated by the project. A state might well prefer this option to closing a toll road facility with anticipated future revenues.

State Infrastructure Banks. The National Highway System Act of 1995 (NHS) created the SIB pilot program to allow states to use federal funds to capitalize a leveraged revolving loan fund for transportation. This program will facilitate a permanent, self-renewable source of capital dedicated to transportation infrastructure. Ten states (Arizona, California, Florida, Missouri, Ohio, Oklahoma, Oregon, South Carolina, Texas and Virginia) have been formally designated as SIB pilot programs and will begin operation during 1997.

There are two types of SIBs: un-leveraged and leveraged. An un-leveraged fund, as provided in the current NHS law, would simply lend funds and provide credit enhancement to eligible applicants. The loan repayments could be used to create the SIB for future transportation projects. Still, the second round of loan recipients would be required to wait a number of years, probably five to seven, until such a fund is created. The un-leveraged SIB rebuilds capital only through repayments from borrowers.

In contrast, a leveraged SIB would issue bonds against its initial capitalization, potentially doubling or tripling the funds available to lend. Instead of directly loaning federal funds to the projects, the funds can be used along with state contributions from non-federal sources to capitalize a leveraged SIB. This could then leverage additional private and public capital by providing credit enhancement, guarantees, collateral, financing or refinancing for qualifying projects, as well as subsidizing the interest rates of projects and creating reserve funds. All these leverage features would expand the funding available for transportation infrastructure investment, accommodating projected capital shortfalls by accelerating the funding of needed projects or reducing the cost of borrowing by improving credit rating.

Creation of the SIBs offers significant opportunities for state DOTs to offer low-cost financing for public-purpose transportation projects sponsored by private partners. This assistance could be in the form of credit enhancement, construction loans or insurance against early project risks. The form and terms of the assistance offered by SIBs will be determined by each state, but federal law and policy permit all forms of financial assistance to be offered to private as well as public project sponsors.

Innovative Income Generation

Under the flexibility permitted by ISTEA and FHWA's TE-045, states may engage in income-generating commercial activities on federal-aid highways. For example, under TE-045, a number of states have proposed commercial operation of rest areas and visitor centers. These public–private partnerships would provide revenue in the form of annual lease payments, as well as create more convenience for the traveling public.

A special subset of these income-generating arrangements relates to telecommunications infrastructure. Under these "shared resources" arrangements, the private-sector partner gains access to public ROW, and the public partner gains access to some form of compensation, either in-kind telecommunications facilities or service, cash, or both. Several states have already entered into such arrangements.

For example, the Maryland Department of General Services has a shared resource agreement with MCI and Teleport Communications Group for the installation of 75 miles of fiber optics along I-95. Maryland will receive access to fiber-optic equipment, as well as maintenance services. The Ohio Turnpike Commission has several nonexclusive licensing agreements with private firms for installing telecommunications infrastructure along ROW, in exchange for a fixed annual license fee of $1,600 per mile and rights to use the fiber optics for turnpike purposes at low or no cost. These public–private partnerships were enabled by FHWA delegation of authority to states to determine their own policies on this matter.

Future Developments in Highway Finance

In addition to the opportunities created by existing federal policy and legislation, new opportunities may be created by upcoming and potential legislative initiatives, especially given that ISTEA's original six-year authorization expired at the end of September 1997. These opportunities include:

- Providing access to un-obligated balances.
- Creating an insurance scheme to protect the private sector from the risks of entering into public–private arrangements.
- Relaxing tax-policy constraints to permit private access to tax-exempt capital for public-purpose projects.

Provide Access to Unobligated Balances

A number of states are seeking access to so-called "unobligated balances" of federal funds. These balances represent the difference between the amounts that a state is apportioned, and the annual amount it is permitted to obligate for contracts. By controlling obligations annually, Congress exercises greater control over spending in response to budget conditions, and uses trust fund revenues for deficit reduction. These balances have grown over the years. California has an accumulated unobligated balance of $650 million; Texas, $745 million; and Massachusetts, nearly $800 million. Currently, total unobligated balances are estimated at $10 billion.

Federal and state policy makers are currently exploring ways to "monetize" these obligations with minimal impact on the federal budget, and are seeking to develop alternatives that would provide less direct, yet still valuable, assistance to financing surface transportation at limited budget cost or risk to the federal government. One possibility would be to allow state DOTs—either with or without SIBs—to use unobligated balances as credit enhancement for public–private projects. This would ensure these funds are used for their original purpose of supporting highway investment, without significantly affecting deficit reduction goals.

Create a Domestic Scheme

In addition to the credit enhancement available from SIBs and Section 1012, the federal government may wish to consider using unobligated balances or other sources of available revenue to create a domestic insurance scheme specifically designed to insure against the risks associated with public–private arrangements.

The Overseas Private Investment Corporation (OPIC) could serve as the model for such a scheme. OPIC insures U.S. companies for risks that traditional insurance is not available to cover, or is prohibitively expensive. These could include political risks, risks of economic downturn, the risk that development does not occur, or the risk that construction will not be successfully completed. This insurance could be offered to prospective private partners at the outset, to ensure that their development costs will not be lost. For example, if the proposers who unsuccessfully participated in the State of Washington's public–private partnerships initiative had been able to insure against political risk using a domestic insurance scheme, they could have recovered their development costs. Without such insurance, future proposers may be reluctant to invest in developing proposals.

Relax Tax Policy Constraints

Many public finance professionals and state and local government officials agree that the provisions of the tax code go far beyond what is necessary to achieve its purposes. The provisions make tax-exempt financing difficult or impossible for activities that are clearly public in nature but for which some measure of private involvement has led to their designation as "private" activities, despite their public purpose. If, for example, a municipality plans a toll facility that serves the general public and is necessary, proponents of tax-exempt debt argue it should be considered a government or public activity regardless of how the facility is financed. They argue the "private" activity bond distinction ignores this fact and classifies some bonds for activities that not only clearly serve but are absolutely necessary to the general public in the same restrictive manner as those that serve only a limited segment of the population. Relaxing this constraint would remove an important barrier to financing public–private projects.

Transportation policies, ISTEA, the NHS Act and the coming ISTEA reauthorization provide incentives for more extensive private-sector involvement in highway development and operation. At the same time, remaining financial barriers still discourage private investment in highway facilities. To build state and local government capacity, and to attract private-sector capital to finance highway investments, the federal government should examine these barriers and consider allowing states to access unobligated balances for credit enhancement. This would create a domestic insurance scheme for public–private partnership arrangements. The federal government should also consider proposals to reclassify transportation bonds as government bonds when the proceeds are used to finance public-property highway facilities. This represents an opportunity for the administration, Congress and the nation as a whole to raise the level of thinking in terms of investment and engage in an honest dialogue of ends and means.

One thing is certain—these ideas will encounter opposition from those who look to defend the past. This is their métier. However, this new line of financial arrangements can coexist with other financial arrangements. The way these issues are decided will say much about how serious we are about rebuilding America.

Notes

1. David A. Aschauer, "Highway Capacity and Economic Growth," *Economic Perspectives* (Federal Reserve Bank of Chicago) 14, no. 5 (September 1990): 14–24.
2. *Rebuilding America: Partnership for Investment,* a Federal Highway Administration report (Washington, DC: U.S. Department of Transportation, March 1995).

The Case for Intellectual Dishonesty

A cynic might be forgiven for insisting that much is to be said in favor of intellectual dishonesty in a society like the United States. He would remind us that we're much more likely to enjoy an adequate supply of the public goods and services that are so vital to our national welfare if we can convince ourselves that "someone else" is paying for them. Because whenever we must admit to ourselves that the cost is coming out of our own pockets, we inevitably try to cut corners, do things on the cheap and ultimately deprive ourselves of much that we really need.

He would also remind us that government has played a major role in this national con game since the early days of the Republic. By cleverly manipulating tax rates and deductions, budget outlays and inflows, aid distribution formulas, public accounting practices, trust fund investments and purchase contracts, government has made it easy for us to persuade ourselves that "the other guy" is paying most of the bill for the things we need. All of which has helped make America great—in the sense of becoming the world's most ostensibly successful national economy during the second half of the 20th century.

Enter "Privatization"

If our friendly cynic is right, the current debate about "privatization" and the most appropriate role for government in providing public services may be missing the mark. This is because it insists on trying to conduct itself in an intellectually honest fashion.

Adapted from Joseph M. Giglio, "Lies! Why Governments Always *Lie About the Cost of Public Works Projects—and Why People Want Them To,"* The American Outlook, *Winter 1998, 19–26.*

These days, privatization is usually defined as a process through which government arranges to have private firms produce many of the services that have traditionally been produced by public agencies. This has become relatively common in areas such as collecting garbage and parking fines. Some state governments are even experimenting with private operation of their prisons. At the other end of the complexity scale are such ambitious undertakings as the state of California's franchising of a private company to finance, build and operate a new toll freeway. This is the S91 project in Riverside County, which has been operating since late 1995.

In theory, there's no apparent reason why any government service could not be privatized. This extends all the way to such sacred cows as police and fire protection, elementary and secondary education, and even the armed forces. Such presumably boundless potential may be why debates about how far privatization should go have become unnecessarily philosophical, involving stained-glass abstractions such as "public versus private goods," "economic externalities," "natural monopolies," and, of course, "the proper role of government."

When Management is the Key Factor

From one perspective at least, the decision about whether or not to privatize public services almost seems like a no-brainer. This perspective involves the vital role played by sharp management in producing and delivering high-quality services with maximum efficiency. Doing so requires service managers who are bright, well trained and strongly motivated. The most obvious way to attract and hold on to such managers is to pay better than the competition.

But government agencies labor under a serious disadvantage when it comes to attracting good managers. The salaries of elected officials tend to impose an artificial ceiling on how much public agency managers can be paid. Typically, this is much less than their counterparts in the private sector. Even though these ceilings ignore the realities of the marketplace, elected officials are reluctant to raise them by advocating higher salaries for themselves because they believe this looks bad to voters. The result is that too many government agencies lack the management talent they need to deliver services efficiently because they can't pay the going rate for good managers.

The State of New York is a good example. The annual salary of its governor is $130,000. Obviously, the commissioners (CEOs) of the state's operating agencies have to be paid less than the governor. So, for example, the Commissioner of the State's Department of Transportation receives an annual salary of $102,335, which establishes the ceiling for the salaries of the agency's top managers.

But this is an agency with more than 11,000 employees that spends roughly $3.7 billion per year on roads, canals and other transportation facilities throughout the state. It is a larger enterprise than many corporations in the S&P 500.

How many of us would be willing to buy stock in a company this size if it couldn't pay its CEO and other top managers substantially more than $100,000? What level of confidence would we have in the capabilities of such an underpaid management team? And New York State is by no means near the bottom of the list in the salaries of its government agency managers.

To put this issue in perspective, here are the combined 1996 salary and bonus figures for CEOs at various private companies in the transportation business (as published in *Business Week*):

- $560,000 at Burlington Northern Santa Fe
- $567,000 at Southwest Airlines
- $595,000 at Conrail
- $911,000 at Northwest Airlines
- $1 million at Delta Airlines
- $1.3 million at CSX
- $1.5 million at United Airlines
- $1.9 million at Norfolk Southern
- $3.1 million at Union Pacific

These figures do not include stock options and other kinds of deferred compensation that government agencies are unable to offer. Just paychecks.

Private companies argue, with compelling logic, that generous compensation packages for their top managers are an essential requirement for efficient operations. They point out that few professional managers are independently wealthy. With families to support and children to educate, gifted managers have no other choice but to sell their services to the highest bidder. A company that fails to recognize this and act accordingly is scarcely meeting its fiduciary responsibilities to its stockholders.

Why should this logic be any different for government agencies? Many of the basic services they provide are just as necessary to a smoothly functioning economy as the utility, communications, and transportation services delivered by private corporations. Shouldn't we insist that these agencies be managed as well as possible, by the best managers money can buy?

Unfortunately, we Americans seem unable to face this reality. We accept with unquestioning faith the peculiar idea that managers of public agencies should be paid considerably less than their counterparts in the private sector. Thus the nation's best managers flock to higher-paying jobs in private companies because public agencies are unable to compete for them in the marketplace. Which leaves us to complain that our public services too often cost more than they should and are less effective than we need them to be.

Given our peculiarly American attitude toward government and civil servants, there seems little likelihood that these irrational constraints on the salaries of public-sector managers will ever disappear. Thus there may be only one practical solution to the problem of assuring ourselves of good public services at reasonable cost. That is to turn a lot more of them over to private companies who are willing and able to pay top dollar for ace managers.

On its face, this seems like a good argument for having private firms produce virtually all public services. Government's role could then be a purely regulatory one, limited to determining which services the public needs and in what quantity.

SHOULD GOVERNMENT OR THE MARKETPLACE DECIDE?

But why stop there? Is there anything inherent in government that enables it to make the most appropriate decisions about what the public needs? Frederick Hayek and other economists of the free-market "Austrian School" would probably say no. They would point out that the truest voice of the public is heard through the marketplace. Demand for a particular service can only be taken seriously when enough members of the public are willing to pay a fair price for that service. A fair price is defined as one that is high enough to induce private businessmen to produce the service in order to earn profits. All other expressions of demand are frivolous and can be ignored.

To some people, this may suggest that the real debate over privatization isn't about whether government agencies or private firms should produce public services. Rather, it's about whether government or the marketplace should decide which services should be produced.

In theory, homeowners in my community could buy fire protection services from a private firm through annual fees rather than paying taxes to support the town government's fire department. Each homeowner could then decide for himself how much service to buy rather than letting government bureaucrats set the same service standards for everyone.

In practice, of course, we know this would never work. Effective fire protection for any one homeowner in a community requires equally effective fire protection for all homeowners in that community.

Suppose my next-door neighbor were to decide not to buy any fire protection service and his house caught fire while he was away on vacation. There would be nothing to prevent that fire from threatening my house as it burned unchecked, no matter how much fire protection service I might have bought. Fire protection is one of those public services whose proper scope is something the marketplace can't determine on its own. Only government can do this.

Is this so unreasonable? In a democratic society, government is supposed to express the will of the people on a variety of issues. That includes assuring the kind, quality and quantity of public services the voters appear to want. But making sure these services are provided doesn't automatically mean that government must produce them. Not if ways can be found for private companies to produce them at prices the consuming public is willing to pay.

In some cases, this won't turn out to be possible. For example, fire protection service in large cities depends on having an adequate supply of physically robust people who are willing to risk their lives every day and retire with ravaged lungs and hearts that leave them with surprisingly short life expectancies. Imagine the kind of NFL-level salaries a private company might have to offer to attract enough such people. Consider the impact these high labor costs would have on the company's price structure.

Municipal fire departments in urban areas have the advantage of being able to staff themselves at much lower labor costs by tapping the large supply of people for whom a career as a city fireman is a proud family tradition. In cities like New York, what economists like to call the "psychic income" that firemen derive from their work is astonishingly high.

Police protection may be another example of a public service that can't be privatized. Police officers are deputized by government to maintain public order. They're given the authority to use their own judgment and act in what may often seem like arbitrary ways in carrying out this responsibility. They're even allowed to kill people with impunity under certain conditions. It's difficult to imagine the American public ever being comfortable with police officers who were not government employees.

But these examples still leave many public services for which privatization is at least a debatable option.

Classic American Model

Forty years ago, this debate had an interesting twist. Traditional socialists liked to complain that standard American practice was to turn over to the private sector all services that could be produced at a profit. Government was left with the responsibility for providing services that seemed inherently unprofitable but were still judged to be necessary for a smoothly functioning society.

Government met this responsibility in two ways. Sometimes it created agencies to produce these services. In other cases, it subsidized private companies in various direct and indirect ways to produce them. In a narrow sense, government's role was seen as one of helping the marketplace appear to be capable of delivering the public services that society needed.

Of course, traditional socialists didn't believe in the marketplace. They kept arguing that government should be given a fair share of the profitable enterprises—with their profits used to help underwrite the costs of unprofitable enterprises that were regarded as necessary or desirable.

Under this tradition of allowing the private sector to have exclusive domain over profitable enterprises, a uniquely American social compact emerged. Among other things, it left the socialists with no audience for their views about government's proper role in a free and fair society.

This compact encouraged large private companies that produced inherently profitable goods and services to go out of their way to be "good corporate citizens." This meant providing an ever larger number of stable jobs that featured steadily growing purchasing power, generous pensions, comprehensive medical coverage and other fringe benefits; heavily supporting local charities in the cities and towns where they did business; lobbying actively for government programs to address social needs that could not be met by corporate America; and underwriting a broad range of cultural activities. In effect, the Fortune 500 assumed the responsibility for many of the social welfare functions provided by government in western European countries. In so doing, they were willing to accept "adequate profits" in place of "maximum profits."

This social compact began to unravel during the buyout boom of the 1980s. Corporate raiders saw that the profitability of the publicly traded companies they took private could be greatly enhanced by abandoning the corporate citizen role and becoming profit maximizers—with the emphasis on cutting costs. This led to the new culture of employment downsizing, gutting pensions and other employee benefits, moving production to countries with lower labor costs, scrapping support for charities and the arts and so on.

Among other things, this has led to an increasingly drastic separation between the living standards of most American workers and their ultimate bosses. But as the 1996 French parliamentary elections have demonstrated, there's a risk that this can eventually lead society to reject the "new" (actually 19th century) corporate model in favor of a classical European welfare state model which promises higher living standards for "everyone" through tax and income redistribution gimmicks.

But is the welfare state model still viable in a world where capital is much more portable than it has ever been? Presumably, there will always be countries whose strong central governments are willing to impose low living standards on their populations in order to attract new business activity.

Yes, but for how long? The Pacific Rim countries are finding that low living standards can only be imposed for a relatively short time. After that, corporations that are still dependent on low-cost labor must move production elsewhere. This is likely to happen all over the world—even in Africa, which is the last major source of low-cost labor and whose central governments can eventually be expected to impose this model on their vast populations once they've learned how to control them.

In the end, however, growing demands for better living standards throughout the world will have to be met in order to assure the social stability on which profitable business activity depends. The average person is not an economic ideologue. He wants a decent living standard for his family and the prospect of a better life in the future. He's perfectly happy to let the private sector provide this. But if the private sector can't or won't deliver, then he's just as happy to have the whole task (along with the means of production) turned over to government.

But the cost of meeting these demands for rising living standards is high. In the end, it can only be covered by rising levels of profitable business activity. Thus we need to find a suitable model for a working partnership between the public and private sectors that can serve both masters.

Two Contrasting Models

Two very different kinds of partnership models are available to accomplish this. Each country will have to make its own choice between them.

1. The Western European model, with its large central government that provides a wide range of social-welfare services to ensure high living standards for an increasingly educated, increasingly productive and therefore increasingly demanding population. The cost of these services will be covered by relatively high direct and indirect taxes on business activity. Countries adopting this model may become natural homes for those companies that wish to be pure profit-maximizers and concentrate on "the business of business." Such companies will accept high taxes as simply one of the costs of doing business.
2. The former U.S. model, with most of the country's social welfare services being provided directly by private companies (probably as a condition of being allowed to do business in that country). These countries will have much smaller central governments that confine themselves mainly to a regulatory role, which permits relatively low taxes. Companies that are willing to accept "satisfactory profits" and like to be known as good corporate citizens may find their natural homes in these countries.

Theoretically, privatization can play a significant role in both models. Under the Western European model, government can either produce public services itself or contract them out to private firms. Under the former U.S. model, private firms accept the responsibility for producing public services as an implicit part of the cost of doing business. In each case, privatization delegates to government the responsibility for determining the type and quantity of services the public will receive.

THE CATCH

Our friendly cynic is going to insist that the privatization model requires a good deal of intellectual dishonesty if it is to work in a society like the

United States. That's because a great many of our public services actually cost more to produce than we're willing to admit. It is much easier to hide these costs and pretend they don't exist if government continues to produce these services. Here's a simple example.

Suppose a private company sees an attractive business opportunity in building a highway toll bridge to provide transportation service across a wide river. It obtains a franchise from government to build, own, and operate the bridge and charge tolls that are high enough to cover annual costs and generate a fair profit for the company. So the company proceeds to raise the capital needed to construct the bridge by issuing bonds.

When the bridge is ready to open, the company establishes a toll schedule to generate the necessary revenue. This revenue must cover the annual costs of operating and maintaining the bridge and paying debt service on its bonds. But it must also be high enough to cover the cost of something called "depreciation." This is the annual cost of wear and tear on the bridge's physical infrastructure that arises from providing transportation services. It is usually estimated by dividing the bridge's original capital cost by the number of years it can be expected to provide service before requiring major reconstruction.

Even though depreciation is not a cash cost like wages (the actual checks to build the bridge were written during its construction period), it is still a real cost that private sector accounting standards insist must be charged against annual revenues in determining the bridge's profit. The revenue to cover depreciation can then be set aside in a "reconstruction reserve fund" to cover the cost of rebuilding the bridge after it has become worn out through years of heavy use.

Meanwhile, some fifty miles up the river, a public authority issues bonds to build a similar toll bridge. But it charges lower tolls. Why? Because government enterprises in the United States aren't required to recognize depreciation as an annual cost to be charged against revenues. They can simply ignore it, which enables their annual costs to seem lower. In the case of the authority's bridge, this permits what seem like bargain-price tolls compared to the private company's bridge. But when the authority's bridge has worn out, there is no reconstruction reserve fund available to pay for its rebuilding. So the authority must issue new bonds to raise the necessary capital.

The private company issued debt to build its bridge. This debt has been fully paid off by the time the bridge required major reconstruction, and the reserve fund from depreciation was available to cover the costs of reconstruction. Which led to a new depreciation schedule (based on the cost of reconstruction) to assure that money would be available the next time the bridge needed to be rebuilt. For practical purposes, therefore, the company incurred a one-time debt to build a bridge than can presumably last forever.

The public authority also issued debt to build its bridge. This debt had also been fully paid off by the time the bridge needed major reconstruction. But new debt had to be issued to pay for reconstruction because the authority was able to ignore the annual cost of depreciation and so had no reserve fund available. This means that the authority's original debt turns out to be a perpetual debt. This debt is effectively rolled over (plus interest) each time the bridge has to be reconstructed.

The convenient American fiction that government enterprises don't have to bother about depreciation means that the authority's bridge seems "less expensive" for the motoring public to use. Its tolls can be lower because today's motorists don't have to cover the cost of depreciation. That cost is covered in the future with borrowed funds. In effect, we simply pass it on to our children in the form of ever-increasing public debt and larger interest payments.

How Intellectual Dishonesty Works

One definition of intellectual dishonesty is the practice of ignoring reality when it interferes with what we want to believe about the way the world works. This is what government enterprises in the United States do when they pretend that depreciation of capital facilities isn't part of their true annual costs.

But the American public, not its government bureaucrats, is to blame for this. Why? Because Americans insist on receiving more service from government than they're willing to pay for. And they don't ask any questions about the charades that are necessary to sustain this illusion of a free lunch.

How can the privatization model cope with such irrational behavior? The private sector can't be expected to produce services for prices that fail to cover costs. Only the public sector can do this. The poor financial performance of companies that have been involved in some of the early privatization deals shows how grossly we have been underestimating the true costs of providing such public services as elementary and secondary education, prison operations and transportation.

Thus it may turn out that one of government's most important roles under the privatization model is to camouflage the true cost of public services through artful devices that enable each of us to imagine that we're getting more service than we're actually paying for—with some fictional "other guy" covering the difference.

One way to do this is through "build–operate" partnerships. That is, a government agency contracts with a private company to build a facility like a toll bridge and operate it for a period of years equal to its expected life before requiring major reconstruction. But ownership of the facility remains with the government agency, which is therefore responsible for its "ownership costs." One of these ownership costs is depreciation, which the government agency is allowed to ignore. This enables the private company to charge lower tolls since its revenues don't have to cover the cost of depreciation.

At the end of the contract period, the government agency owns a worn-out bridge that must be rebuilt if it is to continue providing transportation services. This capital cost is covered by issuing new government debt—whose interest and principal payments are covered by the government's future tax revenues. The agency can then contract with a private company (possibly the same private company) to rebuild the bridge and operate it for another period of years. And so it goes.

In other words, government has provided motorists with a transportation service they need and want. But it hasn't required them to pay the full cost of this service through user charges. The lower tolls on what everyone imagines to be a self-supporting bridge are possible only because the depreciation portion of its annual costs has been capitalized rather than being covered by toll revenues. Since this debt is paid off with general tax revenues, the end result is a bridge that isn't really self-supporting at all but is partially subsidized by taxes. However, we can all pretend that it's self-supporting and unsubsidized, and that its lower tolls are simply the result of "private sector efficiency." Which makes us feel more comfortable? See how easy it is?

Why Intellectual Dishonesty Works

Economists, accountants and other fiscal moralists may be outraged by this kind of shell game and urge Americans to stop acting like children about these things. But Americans have a long and very pragmatic tradition of believing that fiscal morality, like religion and the law, is a great thing as long as it doesn't get in the way of anything really important. That includes being able to believe that we can get more than we actually have to pay for when it comes to public services.

Is there anything really wrong with capitalizing certain public service operating expenses and passing the debt burden onto our children? Our friendly cynic would respond with a hearty "that depends." Cynics believe that questions of right and wrong should be resolved on the basis of what works and what doesn't, not on the basis of ideology.

For example, fiscal moralists think it's perfectly okay to capitalize a cost that generates a stream of benefits for a number of years rather than only the year in which the cost is incurred. That is why they don't raise their eyebrows when a private company (or a public authority) issues debt to build a toll bridge. The bridge will have many years of life before it requires major reconstruction and will generate toll revenues during each of those years. To a fiscal moralist, that stream of annual toll revenues is the only benefit needed to justify capitalizing the bridge's construction cost.

But toll revenues aren't the only benefits the bridge generates. It also generates benefits for the motorists who use it. The bridge gives them a timesaving way to cross the river (possibly the only practical way to cross the river at that point).

In turn, this can generate even larger benefits for society as a whole. Once motorists have a convenient way to cross the river, they're likely to make more trips across the river. Very few of these trips will be pure joy rides. Nearly all of them will have some economic rationale. Motorists will cross the river to work at jobs on the other side. To buy things on the other side. To sell things on the other side. To deliver things on the other side. Such trips support and generate economic activity. More trips mean more economic activity. More economic activity means a more prosperous society.

Over the long run, economic activity tends to grow at a fairly steady pace. Thus, a larger base volume of economic activity today means a much larger volume twenty-five years from now. Who benefits from that larger volume of economic activity in the future? The next generation. Our children. The very children on whom we may dump a load of debt if we decide to capitalize some of the bridge's annual costs rather than paying for them today "in a fiscally sound manner."

The number of trips that today's motorists will make across that toll bridge depends in part on how much they have to pay for each trip. High tolls (high enough to cover all of the bridge's annual costs, including depreciation) will discourage tripmaking. Lower tolls (lower because we have capitalized some of the bridge's annual costs) will encourage people to make more trips. More trips mean more economic activity. More economic activity today means a more prosperous society in the future.

So what are we really doing when we decide to capitalize some of today's public service costs and thereby pass them on to our children? Our cynical friend would tell us that we're simply making an investment to help ensure a more prosperous society for our children. Those same children will also inherit some of the costs of achieving that more prosperous society. Is that so bad?

If we were intellectually honest, we would make this decision in a thoughtful, open, up-front manner. Weighing the pros and cons. Quantifying the benefits and costs. Conducting elaborate discounted cash-flow analyses. And so on. But if this kind of intellectual honesty is beyond us, we will simply proceed to capitalize some of the annual costs of public services—and pretend that we aren't. Does it really make any difference?

In the last analysis, what matters most is that we get the public services we need in a manner we feel comfortable paying for. Because if we don't feel comfortable with how we pay for these services, we won't have them. And society will be poorer as a consequence.

If this requires capitalizing some of the annual costs of public services so that we dump them onto our children, so be it. Our children will inherit the more prosperous economy that our so-called fiscal sins made possible. Also, they can follow the example of their parents and dump the costs we dumped on them onto their own children.

Maybe our friendly cynic is right. Maybe there really is a great deal to be said in favor of intellectual dishonesty in a society like the United States. In which case, the most important role for government under the privatization model may be to make it possible for Americans to continue playing "Let's Pretend" when it comes to public services.

EPILOGUE

America's Transportation Future

Passage of the Safe, Accountable, Flexible, Efficient Transportation Equity Act: A Legacy for Users (SAFETEA-LU) means that America can now look ahead toward improving our outdated, rapidly deteriorating surface-transportation system. Revolutionary changes in financing, technology and management are needed to keep the American economy from becoming bogged down by last-century inadequacies.

We must integrate what traditionally have been separate travel modes into a single, smoothly functioning system that concentrates on service to the customer rather than on mechanical processes. And we must fund surface transportation in a contemporary, integrated and sufficiently robust manner that recognizes the importance of having individual modes complement and support each other financially.

Existing methods of funding highways and streets have broken down. The American Association of State Highway and Transportation Officials (AASHTO) estimates that $5.3 trillion will be needed during the first quarter of the 21st century to prevent further deterioration of the nation's highways and public transit systems, to overcome the effects of past under-spending, and to keep pace with growing transportation demand. But the present federal motor-vehicle fuel tax and appropriations from state and local government budgets are projected to meet less than two-thirds of these needs.

The solution is to end the practice of regarding access to highways as free. Motorists should be charged reasonable tolls for using highways based on the miles they travel, the kind of vehicles they drive, the

---

Adapted from Joseph M. Giglio, "America's Transportation Future," American Outlook Daily Article *(Hudson Institute), August 26, 2005 (http://www.hudson.org/index.cfm?fuseaction=publication_details=3742).*

amount of pollution they generate, and the levels of traffic demand in effect when they choose to travel. New technology now makes it feasible to retrofit the nation's existing highways to collect such tolls without affecting traffic flow. It eliminates the need for toll booths, land-hungry toll plazas and long lines of motorists queuing up to pay tolls. This makes it possible to deliver highway access to motorist customers through the same kind of marketplace mechanism that we traditionally have used to distribute access to a host of goods and services. Highway tolls should be accompanied by money-back performance guarantees so that each motorist can make meaningful trade-offs between how long a trip takes and how much it costs.

To fund what could be a lengthy transition to a fully user-supported highway network, Congress should enact an immediate (though temporary) increase in the federal motor-vehicle fuel tax. This would be indexed to construction costs in order to attack the backlog of highway repairs and overdue capacity increases that too many decades of revenue-starved underspending have created. Revenue from the fuel tax should be lock-boxed to avoid any questions in the minds of motorists about any of it being "borrowed" by other branches of the federal government to fund non-transportation activities.

As increasing portions of the highway system become retrofitted to generate toll revenue, the fuel tax would be phased out, disappearing entirely when the system becomes completely user supported. But during the interim, the fuel tax would provide the funds needed to restore the highway system and public transportation network to a state of good repair.

Hassle-free toll collection is only one of the benefits that new technology offers the surface transportation system. It also enables us to finance, manage and integrate highways, public transportation and goods movement in new ways that promote safety, economic efficiency, environmental friendliness and social benefits that we couldn't even dream about just a few years ago. Equally important are the many ways that technology can help smooth the flow of people and goods, producing savings in travel time and cost, which are becoming increasingly vital components of economic prosperity.

For example, technology already available or in development can greatly reduce the incidence of motor vehicle accidents on crowded highways and limit personal injuries and property loss in accidents that can't be avoided. This would reduce the increasing financial burden that such events impose on American society and minimize the travel time delays they cause. The same technology also can work around the clock to optimize vehicle spacing, speeds, and routings so that a given number of highway lane miles can move more vehicles in less time.

More effective technology will help break down artificial barriers between surface transportation modes so that people and goods move expeditiously door-to-door through a properly integrated system.

However, superior technology alone is not enough to make a difference. More important is how that technology is managed. Transportation system managers must change their focus from engineering-oriented performance measures, like vehicle-miles-of-travel, to customer-oriented measures that emphasize reduced door-to-door travel time, greater reliability and smoother interfaces between transportation modes.

Most importantly—and most radically—we must convert what is now a disconnected group of separate surface transportation modes into a fully integrated system that travel customers perceive as virtually seamless. A key element is to have the national toll-supported highway network be a major funding source for all surface transportation modes, not just for roadways. New emerging technology enables a state or region to manage transportation modes as a portfolio of assets, allocating resources to individual modes. This use of cross-subsidies is consistent with the business portfolio management concept in multi-product corporations where various product lines reinforce each other.

This visionary approach to creating an integrated surface transportation system poses major new financial, technological and management challenges. But an efficient, well-maintained surface transportation system is a vital element of maintaining national economic competitiveness in a world where the value of saving time and money grows by leaps and bounds. Because these radical changes are likely to take a generation to work their way fully through the surface transportation system, the time to begin is now.